T0201974

11

TESI

THESES

tesi di perfezionamento in Chimica sostenuta il 20 luglio 2007

Moreno Lelli
Università di Firenze
Centro Risonanze Magnetiche
Via Luigi Sacconi, 6
50019 Sesto Fiorentino (FI), Italy

Solution Structure and Solution Dynamics in Chiral Ytterbium(III) Complexes

Moreno Lelli

Solution Structure and Solution Dynamics in Chiral Ytterbium(III) Complexes

EDIZIONI
DELLA
NORMALE

ISBN: 978-88-7642-349-9

. . . to my family

Contents

Acknowledgements

I wish to thank all those who collaborated with me during my Ph.D. research: Prof. Lorenzo Di Bari and Prof. Piero Salvadori who supervised me, and the colleagues of the laboratory: Silvia Ripoli, Guido Pintacuda, Gennaro Pescitelli, Chiara Pastore, and Prof.ssa Anna Maria Caporusso.

Silvia Ripoli deserves my special thanks for her suggestions in the scientific work and, mostly important, her support during these years.

My genuine thanks go, of course, to my family: my father Pier Luigi, my mother Luisa, my sister Susy, and my brother Maurizio. They sustained me in every moment of my studies, and it is essentially merit of them, and in particular my brother, if I have developed a scientific spirit and a passion for Science.

Finally, I would like to thank all those who had a positive impact both on this research and generally in my scientific development.

Thank you very much!

The work of this Ph.D. thesis was carried out at the *Scuola Normale Superiore* di Pisa, and at the *Department of Chemistry and Industrial Chemistry* of the University of Pisa, and at the *ICCOM-Pisa* of the National Research Council (CNR), during the years 2000, 2001, and 2002.

Introduction

In the last two decades, lanthanide chemistry has received a growing interest from many fields of chemistry, going from catalysis to the investigation of biological macromolecules, to the use as contrast agents in MRI medical applications [1]. Especially for their use in catalysis, lanthanides offer a unique and smooth variation of many physico-chemical properties along the series, rich coordination chemistry with high and variable coordination numbers (CN), which makes them suitable elements to design catalysts, able to fit to several substrates and reactions [2, 3].

For all these applications, the structural investigation becomes an important prerequisite for a more rational development of such systems. For example, in asymmetric catalysis, different lanthanides modulate small structural variations, which play an important role in the stereoselectivity [3].

The research developed in the present thesis concerns the solution study of structure and dynamics of chiral ytterbium complexes. Some of the investigated molecules are recognized asymmetric catalysts, while some other ones have a potential interest in this field and represent systems of widespread interest in the lanthanide chemistry and in its applications.

We focused our study on ytterbium systems, as the Yb(III) ion offers unique spectroscopic properties, with particular reference to the paramagnetic contribution to NMR, and to the optical properties observable through near-infrared circular dichroism (NIR CD).

Indeed, Yb(III) has a $[Xe]4f^{13}$ electronic configuration, which is paramagnetic for one unpaired electron. The paramagnetic interaction between this electron and the NMR sensitive nuclei of the ligand strongly affects the ligand NMR spectra, influencing both shift and relaxation times [4]. The paramagnetic NMR shift results from the sum of two contributions: the contact (or Fermi) shift, arising from the electron delocalisation, and the pseudocontact (or dipolar) shift due to the (long-range) electron-nucleus dipolar interaction [4]. Referring to Chapter 2 for a more detailed discussion of the theoretical bases of this method, we just mention here that the pseudocontact shift is directly related to the nuclear position, and that the analysis of these shifts with a specially designed computer program (PERSEUS), allowed us to optimise a structure of the complex from the experimental data. In general, for lanthanide complexes, the separation between contact and pseudocontact shift is needed to have structural restraints, thus several elegant procedures were developed [5, 6, 7]. In the case of the Yb(III)-complexes, it is commonly accepted that the contact contributions are small, compared to the dipolar shifts, and can be safely neglected for structural inferences, at least for nuclei separated by more than three bonds from ytterbium [8, 9, 10, 11, 12, 13]. The results presented and discussed in this thesis confirmed the validity of this assumption [14].

The program PERSEUS was developed more or less at the same time of the progress of this thesis, which constituted a relevant test and benchmark for the various features of this software.

In addition to the magnetic properties, Yb(III) has peculiar optical properties [15]: its optical spectrum shows $f-f$ *intra*-configurational transitions, whose absorptions fall in the near-IR range within 850-1050 nm [16, 17]. In chiral complexes, the detection of optical transitions is much easier in CD, because it derives only from chiral Yb-species, and is not influenced by contributions due to achiral species or to the solvent. Furthermore, the NIR wavelengths are sufficiently far from the ligand UV absorptions to avoid interferences with the ligand transitions. Sign, intensity and wavelength of the bands reflect the ligand crystal field around the metal and the chiral distortion of the coordination polyhedron [17]. As a result, NIR CD provides a fast and easy technique to monitor structural changes in solution involving the coordination sphere around the metal (as well as ligand exchange, expansion of the coordination number, structural rearrangement, etc ...) In recent years, the progress in the study of NIR CD of Yb-complexes has offered a series of reference spectra of complexes for which the solution structure and the NIR CD spectrum are simultaneously known [13, 14, 18, 19, 20, 21]. This allows one to interpret the NIR CD spectrum in terms of analogies with similar complexes and to make reasonable hypotheses about the geometry of the coordination polyhedron.

The present thesis demonstrates how paramagnetic NMR and NIR CD can be complementarily used for a detailed structural study of Yb-complexes. The former provides a refined solution structure of the complex, the latter information about the chiral arrangement around the metal and allows one to easily investigate dynamic processes as ligand exchange, proton exchange, etc ... not excluding the use of other techniques, as UV CD, ESI MS to get a complete description of the investigated systems.

The final aim of this work extends itself to define a general strategy for the structural study of Yb(III)-complexes and to test how these techniques can provide precious information on both structural and dynamical aspects. For such a reason, we investigated several molecules ranging from the stable polydentate complexes to labile aggregates.

The present thesis is organized in two parts: the first part is composed of Chapters 1, 2, and 3, and concerns the introduction to the experimental techniques as NMR and NIR CD and to the methods used in this work.

Chapter 1 is a brief summary about the physico-chemical properties and the main aspects of lanthanides coordination chemistry.

Chapter 2 is an introductive chapter about the analysis of the paramagnetic contributions to NMR. In the first part of the chapter the contributions of the paramagnetic interaction to shift and relaxation times are described. The second part of the chapter shows how these experimental constraints can be used to optimise a solution structure through the computer program PERSEUS.

Chapter 3 is an introductive chapter on the NIR CD spectroscopy of Yb(III) complexes. Some examples, taken from literature, are used to illustrate how NIR CD can be used to derive structural information on the investigated complex.

The second part is composed of Chapters 4, 5, and 6, and contains the results obtained in this work. In each chapter, a different category of molecules is studied, developing and testing a methodology for the structural study in solution. In order to introduce the reader to the different topics, the first sections of each chapter (namely Sections 4.1, 5.1, 6.1), were dedicated to an introduction to the analysed subjects.

In Chapter 4 we studied the heterobimetallic complexes $Na_3[Yb((S)$-BINOL)_3]$ and $K_3[Yb((S)$-BINOL)_3]$.[1] These are two catalysts, active in the enantioselective hydrophosphonylation of cyclic imines [22], and belonging to a family of heterobimetallic catalysts introduced first by the group of Shibasaki in the 1992 [3, 23]. The study was oriented to the determination of the solution structure, through the analysis of paramagnetic NMR data, which demonstrate the existence of differences from the X-ray crystallographic structure. The solution structure was completed with a study of the ligand lability, investigating the exchange between the bound BINOL in $K_3[Yb((R)$-BINOL)_3]$ with both free (R)-BINOL-H_2

---

[1] BINOL is enantiopure 1,1'-bis(2-naphtholate), and BINOL-H_2 indicates enantiopure binaphthol.

and (S)-BINOL-H$_2$. These studies revealed the labile character of the ligand and shed new light on the behaviour of such systems, which may orient the hypotheses about the catalytic mechanism.

In the Chapter 5 we investigated [Yb((S)-THP)] and [Yb(THED)], two complexes structurally analogue to DOTA (see Figure 5.1 and Appendix B, C).[2] The tetraazamacrocyclic ligands are a family of polydentate ligands able to give very stable chelates with lanthanides (with stability constants up to 10^{28} for DOTA). The stability of the complexes stimulated in the past years a large interest toward these molecules, in particular for their biological application, as they can be used to vehiculate lanthanides in biological tissue (for example [Gd(DOTA)] found widespread application as MRI contrast agent). Furthermore, [Ln(THED)] complexes were found active catalysts in the phosphate ester hydrolysis, with potential application in the RNA cleavage [24, 25]. In spite of these potentialities, the Ln^{3+} complexes of THED and the homochiral (S)-THP were not yet structurally investigated in detail: only two crystallographic structures, corresponding to the heterochiral species [Eu((R,S,R,S)-THP)(H$_2$O)]$^{3+}$ and [Eu((R,R,R,S)-THP)(H$_2$O)]$^{3+}$, have been reported up to now [26]. Thus, their ytterbium complexes [Yb((S)-THP)] and [Yb(THED)] represent an interesting subject for a detailed structural solution study.

We report the solution structure of [Yb((S)-THP)] and [Yb(THED)], discussing conformational dynamics, as well as solvent coordination and the effect of pH. [Yb((S)-THP)] was studied in several solvents observing that in non-protic solvents (as acetonitrile or DMSO), the complex undergoes a dimerization process with a consequent structural rearrangement of the monomeric moieties. The analysis of the NMR spectra allowed us to determine also the solution structure of the dimer.

In Chapter 6 we investigate the formation of labile adducts between Yb-complexes and chiral diols. In particular it was studied the interaction between diols and [Yb(fod)$_3$] (a popular shift reagents), and between diols and ytterbium triflate (Yb(OTf)$_3$). These studies culminate with two new methods for the determination of the diol absolute configuration through NIR CD. NIR CD is sensitive only to the diol/Yb-complex chiral adduct, as both the chiral diols and the achiral Yb-species do not give NIR CD. The study, conducted with a large variety of diols (including *prim-sec*, *sec-sec*, and *sec-ter* diols), demonstrated that the shape of the NIR

---

[2] The ligand THED is 1,4,7,10-tetrakis-(2-hydroxyethyl)-1,4,7,10-tetraazacyclo-dodecane, while (S)-THP, which is the homochiral analogue of THED, is 1,4,7,10-tetrakis-((S)-2-hydroxypropyl)-1,4,7,10-tetraazacyclododecane. The DOTA ligand is 1,4,7,10-tetraazacyclododecane-1,4,7,10-tetraacetic acid.

CD spectrum is directly correlated with the diol absolute configuration. As we observe CD in the NIR region, this method can be applied to the analysis of diols with strong UV absorption, which can not be analysed otherwise [27, 28].

The combination of NIR CD and paramagnetic NMR is thus demonstrated to be particularly effective to investigate conformational and configurational issues, as well as to follow dynamic processes.

References

[1] The whole volume *102* of *Chem. Rev.* **2002**, is entirely dedicated to the recent frontiers in lanthanide chemistry.

[2] (a) KOBAYASHI, S., *Lanthanides: Chemistry and Use in Organic Synthesis, Topics in Organometallic Chemistry*; Springer-Verlag: Berlin, **1999**. (b) EVANS, C. H., *Biochemistry of the Lanthanides*, **1990**, Plenum Press, New York.

[3] SHIBASAKI, M.; YOSHIKAWA, N. *Chem. Rev.* **2002**, *102*, 2187–2209.

[4] (a) BERTINI, I.; LUCHINAT, C., *Coord. Chem. Rev.* **1996**, *150*. (b) BERTINI, I.; LUCHINAT, C.; PARIGI, G., *Solution NMR of Paramagnetic Molecules*, **2001**, Elsevier, Amsterdam. (c) BERTINI, I.; LUCHINAT, C.; PARIGI, G., *Prog. Nucl. Magn. Reson. Spec.* **2002**, *40*, 249–273.

[5] PETERS, J. A.; HUSKENS, J.; RABER, D. J., *Prog. Nucl. Magn. Reson. Spec.* **1996**, *28*, 283–350.

[6] (a) REUBEN, J.; ELGAVISH, G. A., *J. Magn. Reson.* **1980**, *39*, 421–430. (b) REUBEN, J., *J. Magn. Reson.* **1982**, *50*, 233–236. (c) PLATAS, C.; AVECILLA, F.; DE BLAS, A.; GERALDES, C. F. G. C. ; RODRÍGUEZ-BLAS, T.; ADAMS, H.; MAHIA, J., *Inorg. Chem.* **1999**, *38*, 3190–3199. (d) RIGAULT, S.; PIGUET, C., *J. Am. Chem. Soc.* **2000**, *122*, 9304–9305. (e) GERALDES, C. F. G. C.; ZHANG, S.; SHERRY, A. D., *Inorg. Chim. Acta* **2004**, *357*, 381.

[7] For recent review and developments about this subject: (a) PIGUET, C.; GERALDES, C. F. G. C., in *Handbook on the Physics and Chemistry of Rare Earths*; GSCHNEIDNER, K. A. JR.; BÜNZLI, J.-C. G.; PECHARSKY, V. K. (eds.), **2003**, Elsevier: Amsterdam. Vol. 33, p. 353–463. (b) OUALI, N.; RIVERA, J.-P.; CHAPON, D.; DELANGLE, P.; PIGUET, C., *Inorg. Chem.* **2004**, *43*, 1517–1529. (c) TERAZZI, E.; RIVERA, J.-P.; OUALI, N.; PIGUET, C., *Magn. Reson. Chem.* **2006**, *44*, 539–552.

[8] (a) REILLEY, C. N.; GOOD, B. W.; DESREUX, J. F., *Anal. Chem.* **1975**, *47*, 2110. (b) DESREUX, J. F., *Inorg. Chem.* **1980**, *19*, 1319–1324. (c) STAINER, M. V. R.; TAKATS, J., *J. Am. Chem. Soc.* **1983**, *105*, 410. (d) BRITTAIN, H. G.; DESREUX, J. F., *Inorg. Chem.* **1984**, *23*, 4459–4466. (e) SHELLING, J. G.; BJORNSON, M. E.; HODGES, R. S.; TENEJA, A. K.; SYKES, B. D., *J. Magn. Reson.* **1984**, *57*, 99.

[9] KEMPLE, M. D.; RAY, B. D.; LIPKOWITZ, K. B.; PRENDERGAST, F. G.; RAO, B. D. N., *J. Am. Chem. Soc.* **1988**, *110*, 8275–8287.

[10] AIME, S.; BOTTA, M.; ERMONDI, G., *Inorg. Chem.* **1992**, *31*, 4291–4299.

[11] PETERS, J. A.; HUSKENS, J.; RABER, D. J., *Prog. Nucl. Magn. Reson. Spec.* **1996**, *28*, 283–350.

[12] DI BARI, L.; PINTACUDA, G.; SALVADORI, P., *Eur. J. Inorg. Chem.* **2000**, 75–82.

[13] LISOWSKI, J.; RIPOLI, S.; DI BARI, L., *Inorg. Chem.* **2004**, *43*, 1388–1394.

[14] DI BARI, L.; LELLI, M.; PINTACUDA, G.; PESCITELLI, G.; MARCHETTI, F.; SALVADORI, P., *J. Am. Chem. Soc.* **2003**, *125*, 5549–5558.

[15] BÜNZLI, J.-C. G.; PIGUET, C., *Chem. Soc. Rev.* **2005**, *34*, 1048–1077, and references therein.

[16] RICHARDSON, F., *Inorg. Chem.* **1980**, *19*, 2806.

[17] DI BARI, L.; PINTACUDA, G.; SALVADORI, P., *J. Am. Chem. Soc.* **2000**, *122*, 5557–5562.

[18] DI BARI, L.; PINTACUDA, G.; SALVADORI, P.; DICKINS, R. S.; PARKER, D., *J. Am. Chem. Soc.* **2000**, *122*, 9257–9264.

[19] DI BARI, L.; SALVADORI, P., *Coord. Chem. Rev.* **2005**, *249*, 2854–2879.

[20] DI BARI, L.; PESCITELLI, G.; SHERRY, A. D.; WOODS, M., *Inorg. Chem.* **2005**, *44*, 8391–8398.

[21] DICKINS, R. S.; PARKER, D.; BRUCE, J. I.; TOZER, D. J., *Dalton. Trans.* **2003**, 1264–1271.

[22] (a) GRÖGER, H.; SAIDA, Y.; SASAI, H.; YAMAGUCHI, K.; MARTENS, J.; SHIBASAKI, M., *J. Am. Chem. Soc.* **1998**, *120*, 3089–3103. (b) SCHLEMMINGER, I.; SAIDA, Y.; GRÖGER, H.; MAISON, W.; DUROT, N.; SASAI, H.; SHIBASAKI, M.; MARTENS, J., *J. Org. Chem.* **2000**, *65*, 4818–4825.

[23] SASAI, H.; SUZUKI, T.; ARAI, S.; ARAI, T.; SHIBASAKI, M., *J. Am. Chem. Soc.* **1992**, *114*, 4418–4420.

[24] MORROW, J. R.; BUTTREY, L. A.; SHELTON, V. M.; BERBACK, K. A., *J. Am. Chem. Soc.* **1992**, *114*, 1903–1905.

[25] (a) CHIN, K. O. A.; MORROW, J. R., *Inorg. Chem.* **1994**, *33*, 5036–5041. (b) MORROW, J. R., AURES, K., EPSTEIN, D., *J. Chem. Soc., Chem. Commun.* **1995**, 2431–2432. (c) EPSTEIN, D. M.; CHAPPELL, L. L.; KHALILI, H.; SUPKOWSKI, R. M.; HORROCKS, W. D. J.; MORROW, J. R., *Inorg. Chem.* **2000**, *39*, 2130–2134. (d) BAKER, B. F.; KHALILI, H.; WEI, N.; MORROW, J. R., *J. Am. Chem. Soc.* **1997**, *119*, 8749–8755 and reference therein.

[26] CHIN, K. O. A.; MORROW, J. R.; LAKE, C. H.; CHURCHILL, M. R., *Inorg. Chem.* **1994**, *33*, 656–664.

[27] (a) DILLON, J.; NAKANISHI, K., *J. Am. Chem. Soc.* **1975**, *97*, 5417; (b) PARTRIDGE, J. J.; TOOME, V.; USKOKOVIČ, M. R., *J. Am. Chem. Soc.* **1976**, *98*, 3740–3741; (c) PARTRIDGE, J. J.; SHIUEY, S.; CHADHA, N. K.; BAGGIOLINI, E. G.; BLOUNT, J. F.; USKOKOVIČ, M. R., *J. Am. Chem. Soc.* **1981**, *103*, 1253–1255.

[28] (a) DI BARI, L.; PESCITELLI, G.; PRATELLI, C.; PINI, D.; SALVADORI, P., *J. Org. Chem.* **2001**, *66*, 4819–4825. (b) DI BARI, L.; PESCITELLI, G.; SALVADORI, P., *Chem. Eur. J.* **2004**, 1205–1214.

PART I

Chapter 1
Basic aspects of lanthanide chemistry

The rare earth elements are a large subgroup of the periodic table that offers a unique and gradual variation of many chemical and physical properties. Furthermore, the differences with respect to the d-transition metal complexes make these elements attractive for application in catalysis, and organometallic chemistry. In this chapter will briefly summarize the main chemical properties of the lanthanides derived from their peculiar electronic configuration (Section 1.1), and how these properties reflect in the coordinative behaviour of the rare earth complexes (Section 1.2).

1.1. Electronic configuration and chemical properties

The electronic configuration of lanthanides is characterized by the gradual filling of the inner $4f$ orbitals ([Xe]$6s^2 5d^1 4f^n$, where n goes from 0 for La to 14 for Lu). The outer $6s$ and $5d$ electrons have ionisation potentials comparable to those of the alkali and alkali earth metals (Table 1.1), and are easily lost to give cationic species. The trivalent state Ln(III) is generally the most common oxidation state, but tetravalent (Ln(IV)) and divalent states (Ln(II)) are also known; in particular Ce, Pr, Nd, Tb, Dy and Ho may give Ln(IV) compounds, while Sm, Eu, Er, Tm and Yb may exhibit divalent forms [1]. These *non*-trivalent states are easily reduced or oxidized to more stable Ln(III) forms; only Ce^{4+} and Eu^{2+} can give stable complexes in water.[1] For the purposes of this work, we concentrate our attention on the Ln(III) complexes, even if Sm(II) and Ce(IV) compounds are of great interest in the applications for catalysis and organometallic chemistry [2].

Scrolling along the series, we can see that the electronic configuration for a generic Ln(III) cation is [Xe]$4f^n$ and, except La^{3+} and Lu^{3+}, the Ln(III) ions are paramagnetic. Moreover, the ionisation energies of

---

[1] Aqueous solutions of Ce^{4+} are strong oxidant and aqueous solutions of Eu^{2+} are sensitive to the presence of dissolved oxygen.

Table 1.1. General chemical properties of rare earth elements [1].

Element	Symbol	Electronic Configuration		Ionic Radius (Å)[3]		ΣI^* (eV)	Oxidation States
		Atomic (Ln^0)	Ionic (Ln^{3+})	CN = 6	CN = 9		
Lanthanum	La	$[Xe]6s^25d^1$	$[Xe]4f^0$	1.03	1.22	36.2	3
Cerium	Ce	$[Xe]6s^25d^04f^2$	$[Xe]4f^1$	1.01	1.20	36.4	3, 4
Praseodymium	Pr	$[Xe]6s^25d^04f^3$	$[Xe]4f^2$	0.99	1.18	37.55	3, 4
Neodymium	Nd	$[Xe]6s^25d^04f^4$	$[Xe]4f^3$	0.98	1.16	38.4	3, 4
Promethium	Pm	$[Xe]6s^25d^04f^5$	$[Xe]4f^4$	–	–	–	3
Samarium	Sm	$[Xe]6s^25d^04f^6$	$[Xe]4f^5$	0.96	1.13	40.4	2, 3
Europium	Eu	$[Xe]6s^25d^04f^7$	$[Xe]4f^6$	0.95	1.12	41.8	2, 3
Gadolinium	Gd	$[Xe]6s^25d^14f^7$	$[Xe]4f^7$	0.94	1.11	38.8	3
Terbium	Tb	$[Xe]6s^25d^04f^9$	$[Xe]4f^8$	0.92	1.10	39.3	3, 4
Dysprosium	Dy	$[Xe]6s^25d^04f^{10}$	$[Xe]4f^9$	0.91	1.08	40.4	3, 4
Holmium	Ho	$[Xe]6s^25d^04f^{11}$	$[Xe]4f^{10}$	0.90	1.07	40.8	3, 4
Erbium	Er	$[Xe]6s^25d^04f^{12}$	$[Xe]4f^{11}$	0.89	1.06	40.5	2, 3
Thulium	Tm	$[Xe]6s^25d^04f^{13}$	$[Xe]4f^{12}$	0.88	1.05	41.85	2, 3
Ytterbium	Yb	$[Xe]6s^25d^04f^{14}$	$[Xe]4f^{13}$	0.87	1.04	43.5	2, 3
Lutetium	Lu	$[Xe]6s^25d^14f^{14}$	$[Xe]4f^{14}$	0.86	1.03	40.4	3

$^*\Sigma I$ = *Sum of the first three ionization potentials.*

the elements [3], the optical properties [4], and the magnetic moments of numerous complexes [5] prove that the f orbitals are shielded from the ligand effects, in agreement with the inner character of the $4f$ with respect to the $5d$ orbitals (Figure 1.1).

Consequently, the ligand gives only a minimal perturbation of the electronic potential of the f orbitals [2].

The progressive increase of the nuclear charge along the series is responsible for the well-known ionic radius reduction, which is called *lanthanide contraction* [6]. The ionic radius goes from 1.03 Å in La^{3+} to 0.86 Å for Lu^{3+} in six coordinated complexes (CN = 6), and from 1.22 Å to 1.03 Å for CN = 9 (Table 1.1); these values are comparable with the radius of the Ca^{2+} ion [1].

1.2. Coordination chemistry of Ln(III) ions

The electronic configuration of the Ln^{3+} directly reflects in the coordination chemistry of these metals: the inner character of the f orbitals, and consequently the scarce overlap with the ligand orbitals, is the reason of the predominantly ionic nature of the metal-ligand bond [1, 2]. Moeller *et al.* have pointed out that, even in the most stable complexes, the bond strength is of the same order of magnitude as the $Ln^{3+}-H_2O$ interaction [7]. Measurements of bond distances have also confirmed the

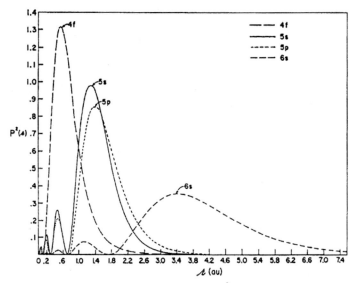

Figure 1.1. Plot of the radial charge density (P^2) of the $4f$-, $5s$-, $5p$-, and $6s$-orbital function for Gd^+ ion (radius in atomic units) [2].

highly ionic nature of lanthanide bonding. Consequently, aspects of the d-transition metal chemistry as the σ donor and π acceptor, the "18 electrons rule", or the formation of multiple bonds ($Ln=O$ or $Ln\equiv N$) are not observed in the lanthanide chemistry.

The electrostatic character of the bond allows variable coordination numbers (CN) and irregular geometries that often can not be simply described as "planar", "tetrahedral" or "octahedral". For this reason, the geometry of the lanthanide complexes is hardly predictable and should be experimentally determined. Furthermore, the ligand arrangement may be different in the solution and the solid-state structure because of the presence of conformational dynamics or of solvent coordination (see examples in Chapter 4 and 5) thus, a specific study is often required. The CN's are usually higher than for d-transition metals and may reach 9 for Yb or Lu and 11 for the larger La [1, 2, 8].

The *non*-directionality of the bond and the CN flexibility are the main aspects of the rare earth coordination chemistry, this makes lanthanides suitable elements in catalysis: potentially, smaller activation energy is required to expand or reduce the coordination sphere of the metal, facilitating the substrate binding and the product release. Furthermore, lanthanide catalysis allows reactions that are "orbitally forbidden" in covalent complexes, offering a valid alternative to the traditional d-transition catalysts.

According the HSAB terminology of Pearson [9], lanthanides cations are usually considered as "hard" Lewis acids, having a strong preference for the oxygen donor atom with respect to the nitrogen and sulphur (O $\gg$ N > S, F > Cl).

Water, alcohols, ethers, DMSO and many common solvents have oxygen donor atoms, so they are coordinating molecules that compete with the lanthanide ligands. Thus, the choice of the solvent is important and severely restricts the possible ligands used, especially when the solvent is water.

Chelation is the predominant form of complexation in lanthanide chemistry, because the entropic driving force, that arises from the coordination of polydentate ligands, compensates for the competition of the solvent and allows the formation of stable complexes. The stability depends on the nature and the number of coordinating donors: chelating N-ligands as ethylendiamine dissociate in water and in oxygenated solvents [1], but if the ligand has O donors, as in EDTA, the stability of the complex greatly increases, with binding constant in water as high as 10^{15}-10^{20} M^{-1}.

The crystallographic structure of the [La(EDTA)(H$_2$O)$_3$]$^-$ shows that three water molecules are bound with an overall CN = 9. The negatively charged carboxylic oxygens are more strongly bound to the metal than the neutral water and the nitrogen, as indicated by the shorter crystallographic bond lengths: 2.54 Å for La$-$O of EDTA against 2.60 Å for La$-$O of water, and 2.86 Å for La$-$N.

Other similar polydentate ligands were derived by changing the number of pendant arms: HEDTA (N'-hydroxyethylethylenediamine-N,N,N'-triacetic acid), DTPA (diethylenetriamine-N,N,N',N',N''-pentaacetic acid) DCTA (1,2-diaminocyclohexane-N,N,N',N'-tetraacetic acid). In particular, DTPA (octadentate with five carboxylic donor) establishes high stability constants in water with all Ln^{3+} ions ($\approx 10^{22}$): for this reason it is used to remove radioactive rare earth metals from the body following accidental exposure [1]. Other macrocyclic polydentate ligands derived from DOTA (1,4,7,10-tetraazacyclododecane-1,4,7,10-tetraacetic acid) are often used in the application and will be described in detail in Chapter 5.

Water is a very strong ligand for Ln^{3+} ions, and it is even observed in the crystallographic structure of complexes that are crystallized from solvent having only small traces of water [10]. In view of the difficulty that competing ligands experience in attempting to replace water, lanthanide complexes are generally synthesized starting from anhydrous salts (as LnCl$_3$, Ln(OTf)$_3$, alcoholates, ...) where the anions are more easily replaceable.

In aqueous solutions, Ln^{3+} ions bind several water molecules: their exact number may be controversial, luminescence measurements have suggested values of 10 for $La^{3+}-Nd^{3+}$, and 9 for $Tb^{3+}-Lu^{3+}$ [11]. The rate constants for the exchange processes in the Ln^{3+} hydration sphere (1.1) are approximately estimated to $8 \cdot 10^7$ s^{-1} for $La^{3+}-Eu^{3+}$, $4 \cdot 10^7$ s^{-1} for Gd^{3+} and $1 \cdot 10^7$ s^{-1} for $Dy^{3+}-Lu^{3+}$, and depend either on the nuclear charge or on the number of water molecules in the coordination sphere

$$[Ln(H_2O)_n]^{3+} + H_2O^* \rightleftharpoons [Ln(H_2O)_{n-1}(H_2O)^*]^{3+} + H_2O \qquad (1.1)$$

$$[Ln(H_2O)_n]^{3+} \rightleftharpoons [Ln(H_2O)_{n-1}(OH)]^{2+} + H^+ \qquad (1.2)$$

Increasing the pH, hydrolysis of bound water is favoured (1.2) leading to the complete precipitation of the lanthanide hydroxides $Ln(OH)_3$. Together with hydroxide precipitation, the formation of polymetallic species as $Ln[Ln(OH)_3]^{3+}$ or colloids may be observed: if the lanthanide hydroxides is desired, a careful control of the precipitation conditions is needed [12]. The $Ln(OH)_3$ solubility decreases along the series in correspondence with the ionic radius reduction, passing from a solubility of $8 \cdot 10^{-6}$ mol/L for La^{3+} to $6 \cdot 10^{-7}$ mol/L for Lu^{3+}.

References

[1] EVANS, C. H., *Biochemistry of the Lanthanides,* **1990,** Plenum Press: New York.

[2] KOBAYASHI, S., *Lanthanides: Chemistry and Use in Organic Synthesis, Topics in Organometallic Chemistry,* **1999,** Springer-Verlag: Berlin.

[3] BÜNZLI, J.-C. G.; CHOPPIN, C. R., *Lanthanides probes in life, chemical and earth science, theory and practice,* **1989,** Elsevier: Amsterdam.

[4] (a) PEACOCK, R. D., *Struct. Bonding* **1975,** *22,* 83. (b) CARNALL, W. T., In: Gschneidner, K. A. Jr.; Eyring, L. *Handbook on the physics and chemistry of the rare earths,* **1979,** North-Holland Publishing Company: Amsterdam, (Chapter 24).

[5] MOELLER, T. IN: MOELLER, T.; SCHLEITZER-RUST, E., *Gmelin handbook of inorganic chemistry, Sc, Y, La-Lu rare elements,* **1980,** part D1, 8th ed.; Springer: Berlin, Heidelberg, New York, (Chapter 1).

[6] QUADRELLI, E. A., *Inorg. Chem.* **2002,** *41,* 167–169 and ref. therein.

[7] MOELLER, T.; MARTIN, D. F.; THOMPSON, L. C.; FERRUS, R.; FEISTEL, G. R.; RANDALL, W. J., *Chem. Rev.* **1965,** *65,* 1–50.

[8] LISOWSKI, J.; STARYNOWICZ, P., *Polyhedron* **2000,** *19,* 465–469.

[9] (a) PEARSON, R. G., *J. Am. Chem. Soc.* **1963,** *85,* 3533. (b) PEARSON, R. G., *Hard and Soft Acids and Bases,* **1973,** Dowden, Hutchinson, and Ross: Stroudsburg, PA.

[10] SASAI, H.; SUZUKI, T.; ITOH, N.; TANAKA, K.; DATE, T.; OKAMURA, K.; SHIBASAKI, M., *J. Am. Chem. Soc.* **1993**, *155*, 10372–10373.

[11] HORROCKS, W. DEW.; SUDNICK, D. R., *J. Am. Chem. Soc.* **1979**, *101*, 334–340.

[12] WITHEY, R. J., *Can. J. Chem.* **1969**, *47*, 4383–4387.

Chapter 2
NMR of lanthanide complexes

Most of the structural results presented in this thesis are obtained from the analysis of NMR data, and especially from the analysis of the paramagnetic contributions to the NMR spectra of lanthanide complexes. Apart for lanthanum (La^{3+}) and lutetium (Lu^{3+}), which have the f-shell completely empty ($4f^0$, for La^{3+}) and full ($4f^{14}$, for Lu^{3+}), in all other Ln^{3+} ions the partial filling of the f-orbitals gives a variable number of unpaired electrons, which are responsible for the molecular paramagnetism.

Irrespective of its origin, molecular paramagnetism directly influences the NMR spectrum, in fact the magnetic moment of the unpaired electrons generates a magnetic field that may interact with the nuclear magnetic dipole (the so called **hyperfine interaction**), and produces additional contributions to both the NMR shift and the relaxation times of the observed nuclei. These paramagnetic contributions (δ^{para}, R^{para}) are determined by taking the difference between the observed shift and relaxation rate (δ^{obs}, R^{obs}) and those expected for same nucleus in a diamagnetic complex (δ^{dia}, R^{dia}) [1]

$$\delta^{obs} = \delta^{dia} + \delta^{para}$$
$$\delta^{para} = \delta^{obs} - \delta^{dia} \tag{2.1}$$

$$R_1^{obs} = \frac{1}{T_1^{obs}} = \frac{1}{T_1^{dia}} + \frac{1}{T_1^{para}} = R_1^{dia} + R_1^{para}$$

$$R_2^{obs} = \frac{1}{T_2^{obs}} = \frac{1}{T_2^{dia}} + \frac{1}{T_2^{para}} = R_2^{dia} + R_2^{para}. \tag{2.2}$$

In this chapter we report a brief description of the main aspects of paramagnetic NMR, with particular regard to the case of lanthanide ions and referring to the literature for a more exhaustive treatment [1, 2, 3].

---

[1] The correct diamagnetic reference is not a trivial matter. In most cases the lanthanum complexes are good references for the early lanthanides (Ce-Eu) and the lutetium complexes are for the late series (Tb-Yb). In the applications described in Chapter 4, 5, 6, each specific case will be discussed.

2.1. The paramagnetic shift

The paramagnetic shift due to the hyperfine interaction can be distinguished in two terms, the **Fermi** or **contact coupling** and the **dipolar** or **pseudocontact coupling**: the first is related to the electron-nucleus magnetic interaction when the electron is localized exactly *on* the nucleus, the second is described by the *long-range* magnetic dipolar coupling. In the nuclear spin Hamiltonian (H_{spin}^{nuc}), contact and pseudocontact couplings provide two terms (H_{con} and H_{pc}, respectively) in addition to the Zeeman (H_{Zeeman}) and chemical shift Hamiltonians (H_{CS})[2]

$$H_{spin}^{nuc} = H_{Zeeman} + H_{CS} + H_{con} + H_{pc}. \tag{2.3}$$

Although both terms accounting for the hyperfine interaction describe a spin-spin coupling, they are both proportional to the electron spin polarization and consequently proportional to the external magnetic field (see details in Section 2.1.1 and 2.1.2). This implies that they give rise to two additional contributions to the NMR shift: a contact (δ^{con}) and pseudocontact (δ^{pc}) term

$$\delta^{para} = \delta^{con} + \delta^{pc} + \delta^{bulk}. \tag{2.4}$$

The additional term δ^{bulk} accounts for the effects derived from the modified magnetic susceptibility of the whole sample, and it is generally cancelled by using an appropriate internal reference [2]: for sake of simplicity this term will be dropped in the following discussion.

2.1.1. The Fermi contact shift

The contact coupling is proportional to the spin density ρ at zero distance from the nucleus, which in turn is the difference of the electron density of the two spin polarizations $S = 1/2$ and $S = -1/2$ [1]

$$\rho = \left| \Psi_{-1/2}(0) \right|^2 - \left| \Psi_{1/2}(0) \right|^2 \tag{2.5}$$

where $\Psi(0)$ is the MO wavefunction *on* the nucleus. The p, d, f, ... orbitals have vanishing electron density on the nucleus, thus only the s-components of the MO contribute to the contact coupling. The contact Hamiltonian describes the coupling between the nuclear and electronic spin operators $\mathbf{I}$ and $\mathbf{S}$, respectively, through the tensor $\tilde{\mathbf{A}}$,

$$H_{con} = \mathbf{I} \cdot \tilde{\mathbf{A}} \cdot \mathbf{S} \tag{2.6}$$

---

[2] Other Hamiltonian terms, as J-coupling, are omitted because they are not influent at this level.

where,

$$\mathbf{I} = \hat{\mathbf{i}}I_x + \hat{\mathbf{j}}I_y + \hat{\mathbf{z}}I_z$$
$$\mathbf{S} = \hat{\mathbf{i}}S_x + \hat{\mathbf{j}}S_y + \hat{\mathbf{z}}S_z . \tag{2.7}$$

For sufficiently high magnetic field,[3] and by replacing the electron spin term with its expectation value $\langle S_z \rangle$,[4] the contact Hamiltonian becomes [1a,b]

$$H_{con} = AI_z \langle S_z \rangle \tag{2.8}$$

where A is the contact coupling constant

$$A = \frac{\mu_0}{3S} \hbar \gamma_I g_e \mu_B \sum_i \rho_i \tag{2.9}$$

and the sum runs over all the s-components of the MOs of the molecule, $\hbar$ is the Planck constant, μ_0 is the magnetic permeability in vacuum, μ_B is the Bohr magneton, g_e is the Landé coefficient for the free electron, γ_I is the nuclear magnetogyric ratio and S is electron-spin quantum number.

Calculating the energy difference for close nuclear levels ($\Delta M_I = \pm 1$), the NMR contact shift becomes

$$\delta^{con} = \frac{A}{\hbar \gamma_I} \frac{g_e \mu_B S(S+1)}{3kT} \tag{2.10}$$

δ^{con} increases lowering the temperature ($\propto T^{-1}$), as expected for the Curie law.

In close analogy to J coupling, the propagation of the electron spin density from the metal center to the observed nucleus occurs through the magnetic polarization of the contiguous electronic clouds (Figure 2.1), and is enhanced in presence of electron delocalisation. In such a way the contact shift may reach to the s-orbitals of hydrogen or hetero-atoms not directly involved in metal bonding.

The through-bond spin density propagation is sensitive to the conformation of the ligand backbone and sometimes it allows one to get structural information. For example, in ligands with nitrogen donors the contact shift may be correlated with the M−N−C−H dihedral θ (Figure 2.2), [1] through a relation analogous to the Karplus rule

$$\delta^{con} = a \cos^2 \theta + b \cos \theta + c . \tag{2.11}$$

---

[3] This condition is satisfied when, for any a_k eigenvalue of $\tilde{\mathbf{A}}$, is $a_k \ll g_e \mu_B B_0$.

[4] This is correct in view of the faster electronic relaxation with respect to the NMR time scale.

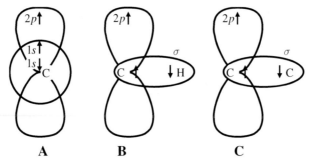

A　　　　　　**B**　　　　　　**C**

Figure 2.1. Propagation of the electron spin density ρ through polarization of the electronic clouds. An unpaired electron, located in a $2p$ orbital, may polarize the electronic clouds in s or σ orbitals, propagating the spin density ρ onto the nucleus (picture A) or onto the closest nuclei (picture B) and (picture C).

This simple expression becomes more complicate changing the donors, the metal, or the ligand complexity.[5] As a consequence, apart from few a cases, obtaining structural information from contact shifts remains a hard matter.

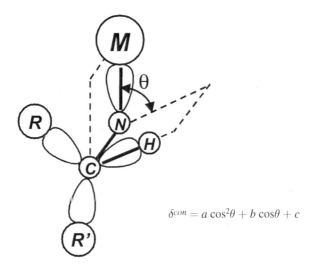

$$\delta^{con} = a\cos^2\theta + b\cos\theta + c$$

Figure 2.2. Conformational dependence of the proton contact shift in proximity of the metal (M).

---

[5] In case of d-transition metal, where the metal coordination may involve s-donation and p-retrodonation, this Karplus-type rule may contain also $\sin^2\theta$ term or more complicated expressions (Ref. [1] and reference therein).

2.1.2. The pseudocontact shift

The pseudocontact coupling is due to the through space *dipolar interaction* between the electron and nuclear magnetic dipoles. In the classical picture, this is described by the dipole-dipole interaction (2.12)

$$E^{\text{dip}} = -\frac{\mu_0}{4\pi} \left[\frac{3(\boldsymbol{\mu}_1 \cdot \mathbf{r})(\boldsymbol{\mu}_2 \cdot \mathbf{r})}{r^5} - \frac{(\boldsymbol{\mu}_1 \cdot \boldsymbol{\mu}_2)}{r^3} \right] \qquad (2.12)$$

where, $\boldsymbol{\mu}_1$ and $\boldsymbol{\mu}_2$ are the nuclear and electronic dipoles, and $\mathbf{r}$ is the vector between them. In the quantum mechanics, the dipolar Hamiltonian is introduced

$$H_{\text{dip}} = -\frac{\mu_0}{4\pi} \hbar \gamma_I \left[\frac{3(\boldsymbol{\mu} \cdot \mathbf{r})\mathbf{r}}{r^5} - \frac{\boldsymbol{\mu}}{r^3} \right] \cdot \mathbf{I} = -\frac{\mu_0}{4\pi} \hbar \gamma_I \boldsymbol{\mu} \cdot \tilde{\mathbf{D}} \cdot \mathbf{I} \qquad (2.13)$$

with $\mathbf{I}$ the nuclear spin operator (2.7), $\boldsymbol{\mu}$ the effective magnetic dipole resulting from the electron distribution and $\tilde{\mathbf{D}}$ the dipolar tensor

$$\tilde{\mathbf{D}} = \frac{1}{r^5} \begin{bmatrix} 3x^2 - r^2 & 3xy & 3xz \\ 3xy & 3y^2 - r^2 & 3yz \\ 3xz & 3yz & 3z^2 - r^2 \end{bmatrix}. \qquad (2.14)$$

For nuclei sufficiently far from the metal the electronic distribution can be approximated to an electric dipole centered *on* the metal (*Metal-centered point-dipole approximation*), and $\tilde{\mathbf{D}}$ contains the coordinates of the nucleus with respect to it. Expressing the magnetic dipole $\boldsymbol{\mu}$ in terms of the external field $\mathbf{B}_0$ and of the magnetic susceptibility tensor $\tilde{\chi}$

$$\boldsymbol{\mu} = \frac{1}{\mu_0} \tilde{\chi} \cdot \mathbf{B}_0 \qquad (2.15)$$

and replacing (2.15) in (2.13), we obtain [12]

$$H_{\text{dip}} = -\frac{\hbar \gamma_I}{4\pi} \mathbf{B}_0 \cdot (\tilde{\chi} \tilde{\mathbf{D}}) \cdot \mathbf{I}. \qquad (2.16)$$

This is formally analogue to the diamagnetic term describing the sum of Zeeman and chemical shift Hamiltonians

$$H_{\text{dia}} = H_{\text{Zeeman}} + H_{\text{CS}} = -\hbar \gamma_I \mathbf{B}_0 \cdot (1 - \tilde{\sigma}) \cdot \mathbf{I} \qquad (2.17)$$

where $\tilde{\sigma}$ is the chemical shielding tensor and $\mathbf{1}$ is the identity operator. Neglecting the H_{con}, the nuclear-spin Hamiltonian (2.3) can be re-written

$$H_{\text{spin}}^{\text{nuc}} = -\hbar \gamma_I \mathbf{B}_0 \cdot \left(1 - \tilde{\sigma} + \frac{1}{4\pi} \tilde{\chi} \tilde{\mathbf{D}} \right) \cdot \mathbf{I}. \qquad (2.18)$$

For a $\mathbf{B}_0$ aligned along the z-axis, and averaging (2.18) for all molecular orientations, the energy difference between levels with $\Delta M_I = \pm 1$ is

$$\Delta E = \hbar \gamma_I B_0 \left(1 - \sigma^{\text{iso}} + \frac{1}{12\pi} \text{tr} \left\{ \tilde{\chi} \tilde{\mathbf{D}} \right\} \right) \qquad (2.19)$$

where, σ^{iso} is proportional to the trace of $\tilde{\sigma}$ ($\sigma^{\text{iso}} = \frac{1}{3}(\sigma_{xx} + \sigma_{yy} + \sigma_{zz}) = \frac{1}{3}\text{tr}\{\tilde{\sigma}\}$). The chemical shift is obtained dividing (2.19) by the Zeeman energy $\hbar \gamma_I B_0$

$$\delta = \frac{\Delta E}{\hbar \gamma_I B_0} = 1 - \sigma^{\text{iso}} + \frac{1}{12\pi} \text{tr} \left\{ \tilde{\chi} \tilde{\mathbf{D}} \right\}$$

$$\delta^{\text{dia}} = 1 - \sigma^{\text{iso}} \qquad (2.20)$$

$$\delta^{\text{pc}} = \frac{1}{12\pi} \text{tr} \left\{ \tilde{\chi} \tilde{\mathbf{D}} \right\}$$

where we made explicit the diamagnetic and pseudocontact shifts. As $\tilde{\mathbf{D}}$ is a traceless tensor, the isotropic component $\chi^{\text{iso}} \mathbf{1}$ of the magnetic susceptibility gives no contribution to δ^{pc}, which depends only on the anisotropic part of the magnetic susceptibility $\tilde{\chi}'$

$$\tilde{\chi}' = \tilde{\chi} - \text{tr}\{\tilde{\chi}\} \mathbf{1} = \tilde{\chi} - \chi^{\text{iso}} \mathbf{1} \qquad (2.21)$$

$$\begin{aligned}
\delta^{\text{pc}} &= \frac{1}{12\pi} \text{tr} \left\{ \tilde{\chi} \tilde{\mathbf{D}} \right\} \\
&= \frac{1}{12\pi} \text{tr} \left\{ (\tilde{\chi}' + \chi^{\text{iso}} \mathbf{1}) \tilde{\mathbf{D}} \right\} \\
&= \frac{1}{12\pi} \left[\text{tr} \left\{ \tilde{\chi} \cdot \tilde{\mathbf{D}} \right\} + \chi^{\text{iso}} \text{tr} \left\{ \tilde{\mathbf{D}} \right\} \right] \\
&= \frac{1}{12\pi} \text{tr} \left\{ \tilde{\chi}' \tilde{\mathbf{D}} \right\}.
\end{aligned} \qquad (2.22)$$

In the metal-centered Cartesian coordinates (2.22) becomes

$$\delta^{\text{pc}} = \frac{1}{12\pi} \frac{1}{r^5} \left[\begin{array}{l} \chi'_{xx} \left(3x^2 - r^2 \right) + \chi'_{yy} \left(3y^2 - r^2 \right) + \chi'_{zz} \left(3z^2 - r^2 \right) \\ + 6\chi'_{xy} xy + 6\chi'_{xz} xz + 6\chi'_{yz} yz \end{array} \right]. \qquad (2.23)$$

By choosing the coordinate system oriented along the principal axes of $\tilde{\chi}'$, this is re-written in (2.24)

$$\delta^{\text{pc}} = D_1 \frac{(3\cos^2 \theta - 1)}{r^3} + D_2 \frac{\sin^2 \theta \cos 2\varphi}{r^3}$$

$$D_1 = \frac{1}{12\pi} \left[\chi'_{zz} - \frac{\chi'_{xx} + \chi'_{yy}}{2} \right] = \frac{1}{12\pi} \left[\frac{3}{2} \chi'_{zz} \right] \qquad (2.24)$$

$$D_2 = \frac{1}{12\pi} \left[\frac{3}{2} (\chi'_{xx} - \chi'_{yy}) \right]$$

where the position of the nucleus is described by the polar coordinates (r, θ, φ in Figure 2.3), and the coefficients D_1 and D_2 contain the principal value of the magnetic anisotropy tensor. For molecules with axial symmetry (C_n with $n \geq 3$), $\chi'_{xx} = \chi'_{yy} = \chi'_\perp$, and $\chi'_{zz} = \chi'_\parallel$, equation (2.24) simplifies to

$$\delta^{pc} = \mathcal{D} \left[\frac{(3\cos^2\theta - 1)}{r^3} \right]$$

$$\mathcal{D} = \frac{\chi'_\parallel - \chi'_\perp}{12\pi} .$$

(2.25)

In both (2.24) and (2.25) the terms $(3\cos^2\theta - 1)/r^3$ and $(\sin^2\theta \cos 2\varphi)/r^3$ explicitly contain the coordinates of the observed nucleus, and make the pseudocontact shifts a precious source of information about the complex geometry (these terms are called *geometrical factors (GF)*). It should be noticed that the pseudocontact shift does not depend on the magnetogyric ratio γ_I and so is independent of the nature of the observed nucleus.

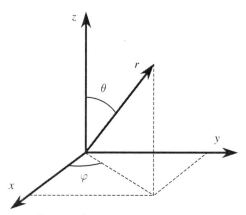

Figure 2.3. Polar coordinate reference system.

In Figure 2.4**A** we report the angular dependence of the pseudocontact shift for tensors with axial symmetry: the plotted surface is symmetric with respect to the equatorial xy plane, and considering the upper hemisphere, all the nuclei that are in the polar region, *i.e.* with $\theta < 54.7°$ are downfield shifted ($\delta^{pc} > 0$) for $\mathcal{D} > 0$ and upfield shifted for $\mathcal{D} < 0$. Conversely, the nuclei in the equatorial region ($\theta > 54.7°$) are upfield shifted ($\delta^{pc} < 0$) for $\mathcal{D} > 0$, and downfield shifted for $\mathcal{D} < 0$. The nuclei at the magic angle ($\theta = 54.7°$) are not shifted by the pseudocontact interaction ($\delta^{pc} = 0$), even when are very close to the metal. For complexes with non-axial tensor, the rhombic term $(\sin^2\theta \cos 2\varphi)/r^3$ must be considered, and the δ^{pc} loses the cylindrical symmetry (Figure 2.4**B,C**).

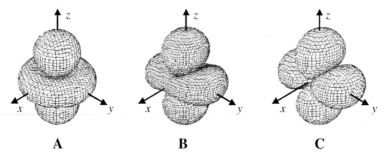

A   B   C

Figure 2.4. Graphical representation of the pseudocontact-shift function. The surfaces are calculated from (2.24) for points having shift of the same magnitude: blue surface indicate negative shift, red surface indicates positive shift. **A** axial symmetry ($D_1 = 1000$; $D_2 = 0$), **B** and **C** non axial symmetry, $D_1 = 2000$; $D_2 = 1000$ and $D_1 = 1000$; $D_2 = 1000$, respectively.

Equations (2.23-25) are effective within the limits of the metal-centered point-dipole approximation: if not, the whole electronic orbital distribution should be considered [4]. For d-transition metals significant deviations from the point-dipole approximation are estimated for distances lower than 7 Å from the metal [4]. At variance, in lanthanides the inner character of the f-orbital strongly reduces this deviation: Golding *et al.*, calculate δ^{pc} from the exact charge distribution in f-orbitals, showing that the deviation from the point-dipole approximation is negligible for distances larger than 3-4 Å [4]. In the case of f^{13} configuration (Yb^{3+}) in a tetragonal distortion, the error in the metal-nucleus distance calculated by using the point-dipole limit is of the order of $(0.6–2.4)\cdot10^{-2}$ Å at 2 Å ($< 1.5\%$) and reduces to less than $(0.4–1.5)\cdot10^{-2}$ Å at 3 Å ($< 0.5\%$).

2.2. Paramagnetic relaxation

In addition to contributing to the hyperfine shift, the electronic paramagnetism affects also nuclear relaxation. Longitudinal and transverse relaxation times (T_1 and T_2, respectively) are generally shortened by the interaction with the paramagnetic center (equations 2.2), with relaxation rates that depend on the nature of the metal and on their distance from the observed nucleus.

Three main mechanisms are responsible for paramagnetic nuclear relaxation: the **dipolar**, the **Curie** and the **contact** mechanisms.

2.2.1. The dipolar relaxation mechanism

The electron-nucleus dipolar interaction (2.12-13) depends on the relative position and orientation of the magnetic dipoles, thus, the continuous reorientation of the molecule with respect to the magnetic field due to the molecular tumbling, as well as the change in the electron-spin-state because of the electronic relaxation, generate stochastic fluctuations of this interaction, and cause **dipolar relaxation**. The correlation time τ_C for these fluctuations is a combination of the rotational correlation time[6] τ_R, the electronic correlation time τ_E,[7] and one derived from the chemical exchange τ_M

$$[dipolar\ \tau_C] \qquad \tau_C^{-1} = \tau_R^{-1} + \tau_E^{-1} + \tau_M^{-1}. \qquad (2.26)$$

The relaxation rates are described by the Solomon equations [5, 6, 7]

$$R_1^{dip} = \frac{1}{T_1^{dip}} = \frac{2}{15}\left(\frac{\mu_0}{4\pi}\right)^2 \frac{\gamma_I^2 g_e^2 \mu_B^2 S(S+1)}{r^6}$$
$$\times \left[\frac{\tau_C}{1+(\omega_I - \omega_S)^2 \tau_C^2} + \frac{3\tau_C}{1+\omega_I^2\tau_C^2} + \frac{6\tau_C}{1+(\omega_I+\omega_S)^2 \tau_C^2}\right] \qquad (2.27)$$

$$R_2^{dip} = \frac{1}{T_2^{dip}} = \frac{1}{15}\left(\frac{\mu_0}{4\pi}\right)^2 \frac{\gamma_I^2 g_e^2 \mu_B^2 S(S+1)}{r^6}$$
$$\times \left[4\tau_C + \frac{\tau_C}{1+(\omega_I - \omega_S)^2 \tau_C^2} + \frac{3\tau_C}{1+\omega_I^2\tau_C^2}\right.$$
$$\left. + \frac{6\tau_C}{1+(\omega_I + \omega_S)^2 \tau_C^2} + \frac{6\tau_C}{1+\omega_S^2\tau_C^2}\right] \qquad (2.28)$$

$$R_{1\rho}^{dip} = \frac{1}{T_{1\rho}^{dip}} = \frac{1}{15}\left(\frac{\mu_0}{4\pi}\right)^2 \frac{\gamma_I^2 g_e^2 \mu_B^2 S(S+1)}{r^6}$$
$$\times \left[\frac{4\tau_C}{1+\omega_B^2\tau_C^2} + \frac{\tau_C}{1+(\omega_I - \omega_S)^2 \tau_C^2} + \frac{3\tau_C}{1+\omega_I^2\tau_C^2}\right.$$
$$\left. + \frac{6\tau_C}{1+(\omega_I + \omega_S)^2 \tau_C^2} + \frac{6\tau_C}{1+\omega_S^2\tau_C^2}\right] \qquad (2.29)$$

---

[6] It should be more appropriate to use the term "reorientational" correlation time, because in solution the full molecular rotation is prevented by the frequent hits with the solvent. As "rotational correlation time" is a very commonly used term, in this work we keep this expression.

[7] The electronic correlation time τ_E is related to electronic relaxation. In principle, longitudinal (T_{1E}) and transverse relaxation times (T_{2E}) may be distinguished in $\tau_{E1} = T_{1E}$ and $\tau_{E2} = T_{2E}$. In the present work we consider the fast motion limit for the electron, with $T_{1E} = T_{2E}$, and consequently $\tau_E = T_{1E} = T_{2E}$.

where all the constants are the same reported above, ω_I and ω_S are the nuclear and electronic Larmor frequencies, respectively. The spectral density terms (reported in the square brackets) describe the efficiency of the relaxation mechanism as a function of the Larmor frequencies and the correlation time. The $R_{1\rho}^{dip}$ term is the longitudinal relaxation rate in the rotating frame, and ω_B is the nuclear Larmor frequency upon rotation around the spin-lock radio frequency field $\mathbf{B}_1$. As $|\omega_B^2 \tau_C^2| \ll 1$, $R_{1\rho}^{dip}$ is substantially equal to R_2^{dip} and will be omitted in the following description [1a,b].

In the fast motion limit ($|\omega_I^2 \tau_C^2| \ll 1$ and $|\omega_S^2 \tau_C^2| \ll 1$) we have

$$R_1^{dip} = R_2^{dip} = \frac{4}{3}\left(\frac{\mu_0}{4\pi}\right)^2 \frac{\gamma_I^2 g_e^2 \mu_B^2 S(S+1)}{r^6}\tau_C. \tag{2.30}$$

The constant $\gamma_I^2 g_e^2 \mu_B^2 S(S+1)/r^6$ is proportional to the square of the nuclear-electron dipolar interaction, and is strongly influenced by the metal-nucleus distance (r^{-6}).

2.2.2. The Curie relaxation mechanism

The fluctuation of the electronic magnetic dipole averages around an "effective" magnetic dipole μ arising from the different populations of the electronic states, and proportional to the external field $\mathbf{B}_0$ (2.15). The dipolar interaction between μ and the observed nucleus fluctuates under the effect of molecular tumbling, and is responsible for the **Curie relaxation**. As μ is averaged over the S states, the correlation time here contains only the rotational and exchange terms

$$[Curie\ \tau_C] \qquad \tau_C^{-1} = \tau_R^{-1} + \tau_M^{-1} \tag{2.31}$$

and the relaxation rates are thus expressed [7, 8, 9]

$$R_1^{Curie} = \frac{1}{T_1^{Curie}} = \frac{2}{5}\left(\frac{\mu_0}{4\pi}\right)^2 \frac{\gamma_I^2 g_e^4 \mu_B^4 D_0^2 [S(S+1)]^2}{(3kT)^2 r^6}\frac{3\tau_C}{1+\omega_I^2\tau_C^2} \tag{2.32}$$

$$\begin{aligned} R_2^{Curie} &= \frac{1}{T_2^{Curie}} \\ &= \frac{1}{5}\left(\frac{\mu_0}{4\pi}\right)^2 \frac{\gamma_I^2 g_e^4 \mu_B^4 B_0^2 [S(S+1)]^2}{(3kT)^2 r^6}\left[4\tau_C + \frac{3\tau_C}{1+\omega_I^2\tau_C^2}\right] \end{aligned} \tag{2.33}$$

that, in the fast motion limit ($|\omega_I^2 \tau_C^2| \ll 1$) become

$$R_1^{Curie} = \frac{6}{7}R_2^{Curie} = \frac{6}{5}\left(\frac{\mu_0}{4\pi}\right)^2 \frac{\gamma_I^2 g_e^4 \mu_B^4 B_0^2 [S(S+1)]^2}{(3kT)^2 r^6}\tau_C. \tag{2.34}$$

The Curie relaxation rates R_1 and R_2 are never equal, even in the fast motion limit, and significantly increase with the external magnetic field ($\propto B_0^2$). This relaxation mechanism is strongly temperature dependent: in addition to the explicit T^{-2} term, the temperature largely influences also τ_R. The magnetic field and temperature dependence in the dipolar relaxation is generally smaller and hardly evaluated, as it depends only on the implicit contribution included in τ_C.

2.2.3. The contact relaxation mechanism

The contact interaction also contributes to the nuclear relaxation with a **contact relaxation** term. In this mechanism, the correlation time τ_C depends only on the electronic and the exchange correlation times, as the contact interaction is essentially scalar and scarcely influenced by molecular orientation

$$[contact\ \tau_C] \qquad \tau_C^{-1} = \tau_E^{-1} + \tau_M^{-1}. \qquad (2.35)$$

The expression for R_1 and R_2 are given by the Bloembergen equations [7, 10]

$$R_1^{con} = \frac{2}{3}S(S+1)\left(\frac{A}{\hbar}\right)^2 \frac{\tau_C}{1 + \omega_S^2 \tau_C^2} \qquad (2.36)$$

$$R_2^{con} = \frac{1}{3}S(S+1)\left(\frac{A}{\hbar}\right)^2 \left[\tau_C + \frac{\tau_C}{1 + \omega_S^2 \tau_C^2}\right] \qquad (2.37)$$

where A is the coupling constant (2.9). In extreme narrowing condition ($|\omega_S^2 \tau_C^2| \ll 1$) (2.36) and (2.37) simplify in

$$R_1^{con} = R_2^{con} = \frac{2}{3}S(S+1)\left(\frac{A}{\hbar}\right)^2 \tau_C. \qquad (2.38)$$

Apart from the donor ligand nuclei themselves, the contact relaxation is generally much smaller than the dipolar and Curie terms and often may be neglected [1]. The relative importance of the dipolar and Curie mechanisms is strongly variable with the nature of the metal, the magnetic field and the electronic and rotational correlation times, and must be evaluated for each specific case. For all the relaxation mechanisms reported above, the relaxation rates are proportional to the square of the magnetogyric ratio (γ_I^2): it derives that proton and fluorine are much more affected by relaxation than most heteronuclei (^{13}C, ^{15}N, etc.).

2.3. Paramagnetic properties of lanthanides

2.3.1. Paramagnetic shift

The paramagnetism of the lanthanide ions is due to the partial filling of the $4f$ orbitals. In these elements, the total electronic spin angular moment (S) and orbital angular moments (L) are strongly coupled by the spin-orbit interaction, so their magnetic properties are better described by the total angular moment J ($|J| = |L \pm S|$).

The equations of the paramagnetic shift and relaxation rates are easily derived replacing the spin quantum number S with J, and the corresponding Landé factor g_e with g_J that contains the orbital contribution. Table 2.1 reports, for each Ln^{3+} ion, the spectroscopic states with the calculated values of J and g_J.

Table 2.1. Electronic properties of lanthanide ions [1]. The spectroscopic states $^{2S+1}L_J$ are referred to the ground state. The pseudocontact shifts (δ^{pc}) are calculated for a nucleus with the $GF = 1$ in an axially symmetric tensor. The values $\langle S_Z \rangle_J$ are calculated including the excited states. The relative linewidths are calculated from (2.48-50) using the electronic correlation time reported in [1], and scaled to Yb linewidth 10 Hz.

Ln(III)	$^{2S+1}L_J(2J+1)$	g_J	$\langle S_Z \rangle_J$	δ^{pc} (ppm)	$\delta^{pc}/\langle S_Z \rangle_J$	Relative Linewidth (Hz)
Ce^{+3}	$^2F_{5/2}$ (6)	6/7	0.98	1.6	1.5	2.3
Pr^{+3}	$^3H_4$ (9)	4/5	2.97	2.7	0.9	5.2
Nd^{+3}	$^4I_{9/2}$ (10)	8/11	4.49	1.0	0.2	5.7
Pm^{+3}	$^5I_4$ (9)	3/5	4.01	-0.6	-0.1	2.7
Sm^{+3}	$^6H_{5/2}$ (6)	2/7	-0.06	0.2	0.2	0.2
Eu^{+3}	$^7F_0$ (1)	-	-10.68	-1.0	0.1	-
Gd^{+3}	$^8S_{7/2}$ (8)	2	-31.50	0.0	0.0	$100\text{-}10^4$
Tb^{+3}	$^7F_6$ (13)	3/2	-31.82	20.7	-0.6	109.5
Dy^{+3}	$^6H_{15/2}$ (16)	4/3	-28.54	23.8	-0.8	151.0
Ho^{+3}	$^5I_8$ (17)	5/4	-22.63	9.4	-0.4	149.0
Er^{+3}	$^4I_{15/2}$ (16)	6/5	-15.37	-7.7	0.5	104.3
Tm^{+3}	$^3H_6$ (13)	7/6	-8.21	-12.7	1.5	47.1
Yb^{+3}	$^2F_{7/2}$ (8)	8/7	-2.59	-5.2	2.0	10.0

The hyperfine **contact shift** in lanthanides is less important than in the d-transition metals [1, 2, 11], because of the inner character of the $4f$ orbitals. The propagation of the contact coupling occurs through overlap of the donor orbitals and the outer $6s$ orbitals of the metal, which are in turn polarized by the $4f$ electrons. The contact shift equation (2.10) corrected for the J term is

$$\delta^{con} = \frac{A}{\hbar\gamma_I} \frac{g_J(g_J - 1)\mu_B J(J + 1)}{3kT} = -\frac{A}{\hbar\gamma_I B_0} \langle S_Z \rangle_J \qquad (2.39)$$

where, A is described in (2.8-9) and $\langle S_Z \rangle_J$ is the expectation value of the electronic spin z-component.[8]

The **pseudocontact shift** is due to the anisotropic part of the magnetic susceptibility (2.22), which is in turn originated by the electronic orbital contributions. However, only when the lanthanide ion is immersed into ligand crystal-field, the spherical symmetry around the metal is removed and the magnetic susceptibility anisotropy arises. Bleaney [12] provided an elegant procedure to derive in a general way the δ^{pc} from the ligand crystal-field parameters. The effect of the crystal-field on the electronic spin Hamiltonian H_{spin}^{el} is spanned into a sum of *spin operators* O_k^q of k rank involving the J_x, J_y, J_z components

$$H_{spin}^{el} = \mu_0 g_J \mu_B \mathbf{J} \cdot \mathbf{B_0} + \sum_{k,q} A_k^q \langle r^k \rangle \langle J \|k\| J \rangle O_k^q \qquad (2.40)$$

where: $k = 2, 4, 6$, and $0 \leq q \leq k$, $\mu_0 g_J \mu_B \mathbf{J} \cdot \mathbf{B_0}$ is the Zeeman term, $\mathbf{B_0}$ is the magnetic field vector, $\mathbf{J}$ is total angular moment operator, A_k^q is the energy crystal-field coefficient of order q and rank k, $\langle r^k \rangle$ is the average of the kth power of the electronic radius in the $4f$ orbitals, and $\langle J \|k\| J \rangle$ are numerical factors calculated from the electronic state of each lanthanide ion. From this Hamiltonian the magnetic susceptibility tensor χ is derived, which is inserted in (2.24) to give

$$\delta^{pc} = -\frac{\mu_0}{4\pi} \frac{g_J^2 \mu_B^2 J(J + 1)(2J - 1)(2J + 3)}{60(kT)^2}$$

$$\times \left[D_Z \frac{3\cos^2\theta - 1}{r^3} + (D_X - D_Y) \frac{\sin^2\theta \cos 2\varphi}{r^3} \right] \qquad (2.41)$$

$$D_X = \langle r^2 \rangle \langle J \|\alpha\| J \rangle (A_2^2 - A_2^0)$$
$$D_Y = \langle r^2 \rangle \langle J \|\alpha\| J \rangle (-A_2^2 - A_2^0)$$
$$D_Z = \langle r^2 \rangle \langle J \|\alpha\| J \rangle (2A_2^0)$$

where D_X, D_Y, D_Z contain the crystal-field parameters. Bleaney [12] observed that while the isotropic component of the magnetic susceptibility depends on the temperature with T^{-1} (Curie law), the anisotropic part

---

[8] It should be mentioned that the constant A contains the electronic Landé factor g_e, which is not replaced by g_J.

depends on T^{-n} with ($n \geq 2$). He assumed that the crystal-field splitting energy (ΔE_{CF}) of the ground state multiplet is small compared to kT, so that only crystal-field terms with rank $k = 2$ can be considered, which give a temperature dependence on T^{-2}.[9]

Assuming an isostructural series of axially symmetric complexes and considering a plausible value of $A_2^0\langle r^2 \rangle = +10$ cm^{-1} for all the lanthanides, Bleaney [12] estimated the pseudocontact shift with (2.41) for a nucleus with $(3\cos^2\theta - 1)/r^3 = 1$ (Figure 2.5 and Table 2.1).[10] Notably, the two Tb^{3+} and Dy^{3+} ions give the strongest shifts, of opposite sign with respect to Er^{3+}, Tm^{3+}, and Yb^{3+}.

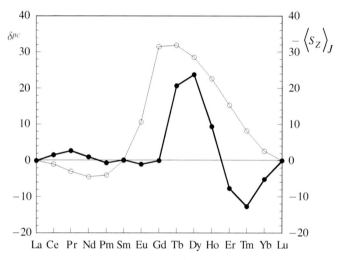

Figure 2.5. Relative contributions of the pseudocontact (δ^{pc}, black points) and contact terms $-\langle S_Z \rangle_J$ (empty points), along the Ln(III) series, as reported in Table 2.1 [1, 12].

In Figure 2.5 these shifts are plotted together with the values of $-\langle S_Z \rangle_J$ (2.39) in order to show the tendency of each metal to give contact and pseudocontact shift contributions.

The T^{-2} dependence of the pseudocontact shift proposed by Bleaney is correct under the assumption that crystal field splitting energy (ΔE_{CF}) of the J manifold is small compared to kT, otherwise the crystal-field

---

[9] The Eu^{3+} and Sm^{3+} ions do not follow (2.41) because of the presence of excited states with J different from the ground state and close in energy to it. These excited states give contributions to the anisotropy of the magnetic susceptibility in T^{-n} with $n > 2$. Anyway, for all the other Ln^{3+} ions these excited states are far in energy from the ground state and these additional terms can be safely neglected (Ref. [12]).

[10] Other values of the crystal field coefficient, even negative, are equally possible; so the data here reported should be considered only for the relative shift intensity in the different lanthanide ions.

parameters of rank $k = 4, 6$ should be considered, which generate additional T^{-n} terms of order higher than 2 ($n > 2$). McGarvey estimated that T^{-3} contributions in the D_1 and D_2 constants (see (2.24)) are not negligible but do not exceed 10% of the T^{-2} term [13]. He concluded that the contact shift (proportional to T^{-1}) can not be easily separated from the pseucontact part (depending on the T^{-2} and T^{-3} terms) by simply fitting the trend of the shifts at variable temperature, because the T^{-1}, T^{-2}, and T^{-3} functions have very similar trends even in a wide range of temperatures [13, 14]. The assumption that the crystal field splitting is smaller compared to kT ($\Delta E_{CF} > kT$) is often not acceptable, as ΔE_{CF} may also be $\Delta E_{CF}/kT > 2$ or higher (see for example Section 3.1.2 and Section 4.5). More recently, Mironov et al. [15] calculated the anisotropy of the magnetic susceptibility D_1 along the lanthanide series, including all the crystal field terms, and for several regular and distorted coordination polyhedrons. These values were compared with those expected from Bleaney's theory, (that limits the crystal field terms to A_2^0). It results that, under a qualitative point of view, the trend of D_1 along the Ln^{3+} series expected from Bleaney (Figure 2.5 and Table 2.1) is still acceptable, but in a quantitative analysis the Bleaney values may show large deviation from the ones calculated including all the crystal field terms. Mironov also evidenced how D_1 has a remarkable dependence on the geometry of the coordination polyhedron and is largely influenced by small distortions in the ligand arrangement [15a].

2.3.2. Paramagnetic relaxation

The paramagnetic relaxation contributions, corrected for the spin-orbit effect, and in absence of chemical exchange, are thus summarized

$$R_1^{\text{dip}} = \frac{2}{15} \left(\frac{\mu_0}{4\pi}\right)^2 \frac{\gamma_I^2 g_J^2 \mu_B^2 J(J+1)}{r^6}$$

$$\times \left[\frac{\tau_C}{1+(\omega_I - \omega_S)^2 \tau_C^2} + \frac{3\tau_C}{1+\omega_I^2 \tau_C^2} + \frac{6\tau_C}{1+(\omega_I + \omega_S)^2 \tau_C^2}\right] \qquad (2.42)$$

$$R_2^{\text{dip}} = \frac{1}{15} \left(\frac{\mu_0}{4\pi}\right)^2 \frac{\gamma_I^2 g_J^2 \mu_B^2 J(J+1)}{r^6}$$

$$\times \left[4\tau_C + \frac{\tau_C}{1+(\omega_I - \omega_S)^2 \tau_C^2} + \frac{3\tau_C}{1+\omega_I^2 \tau_C^2}\right.$$

$$\left. + \frac{6\tau_C}{1+(\omega_I + \omega_S)^2 \tau_C^2} + \frac{6\tau_C}{1+\omega_S^2 \tau_C^2}\right] \qquad (2.43)$$

$$R_1^{\text{con}} = \frac{2}{3} J(J+1) \left(\frac{A}{\hbar}\right)^2 \frac{\tau_E}{1 + \omega_S^2 \tau_E^2} \tag{2.44}$$

$$R_2^{\text{con}} = \frac{1}{3} J(J+1) \left(\frac{A}{\hbar}\right)^2 \left[\tau_E + \frac{\tau_E}{1 + \omega_S^2 \tau_E^2}\right] \tag{2.45}$$

$$R_1^{\text{Curie}} = \frac{2}{5} \left(\frac{\mu_0}{4\pi}\right)^2 \frac{\gamma_I^2 g_J^4 \mu_B^4 B_0^2 [J(J+1)]^2}{(3kT)^2 r^6} \frac{3\tau_R}{1 + \omega_I^2 \tau_R^2} \tag{2.46}$$

$$R_2^{\text{Curie}} = \frac{1}{5} \left(\frac{\mu_0}{4\pi}\right)^2 \frac{\gamma_I^2 g_J^4 \mu_B^4 B_0^2 [J(J+1)]^2}{(3kT)^2 r^6} \left[4\tau_R + \frac{3\tau_R}{1 + \omega_I^2 \tau_R^2}\right]. \tag{2.47}$$

With the exclusion of the donor atoms, the contact relaxation is generally negligible compared to other terms. For all lanthanide ions, apart from Gd^{3+}, the correlation time of the dipolar mechanism is dominated by the electronic correlation time τ_E (10^{-12}-10^{-14}) [1, 16] that is much smaller than the rotational correlation time. The very short value of τ_E is due to the efficient electronic relaxation (Orbach-type process) [17], favoured by the presence of several states near in energy to the ground state. This is not the case of Gd^{3+}, for which the ground state is a singlet ($^8S_{7/2}$) without low lying excited levels, and τ_E is considerably higher (10^{-8}-10^{-9}) [18].

In the absence of chemical exchange (or better for $\tau_E \ll \tau_M$), for small molecules ($|\omega_I^2 \tau_C^2| \ll 1$ and $|\omega_S^2 \tau_C^2| \ll 1$), and for all the lanthanides except Gd^{3+}, equations (2.42)-(2.47) become

$$R_1^{\text{dip}} = R_2^{\text{dip}} = \frac{4}{3} \left(\frac{\mu_0}{4\pi}\right)^2 \frac{\gamma_I^2 g_J^2 \mu_B^2 J(J+1)}{r^6} \tau_E \tag{2.48}$$

$$R_1^{\text{con}} = R_2^{\text{con}} = \frac{2}{3} J(J+1) \left(\frac{A}{\hbar}\right)^2 \tau_E \tag{2.49}$$

$$R_1^{\text{Curie}} = \frac{6}{7} R_2^{\text{Curie}} = \frac{6}{5} \left(\frac{\mu_0}{4\pi}\right)^2 \frac{\gamma_I^2 g_J^2 \mu_B^4 B_0^2 [J(J+1)]^2}{(3kT)^2 r^6} \tau_R. \tag{2.50}$$

2.3.3. Choice of the lanthanide ion

To get structural information from the paramagnetic shifts, the pseudo-contact shift must be separated from the contact shift. As mentioned above, these terms can not be distinguished on the bases of the different temperature dependence [13]. For an isostructural series of lanthanide complexes, the contact and pseudocontact shifts vary in a predictable way, being proportional to $\langle S_z \rangle_J$ and to the Bleaney coefficient ((2.41) and Table 2.1), respectively. The combination of equations (2.3), (2.39), and (2.41) allows one to extract linear plots of the experimental shift as a function of these coefficients, and to isolate contact from pseudocontact contributions [1, 2, 3, 14, 19, 20]. The linearity of these plots may also used to check the isostructurality of the complex geometry along the lanthanide series [2, 3, 20, 21, 22]. Forsberg *et al.* determined the contact and pseudocontact shift for the all the protons of a [Ln(DOTA-like)] derivative analysing the shifts for 10 Ln^{3+} complexes [23]. These data showed that the ytterbium ion has the more favourable ratio between pseudocontact and contact shift, and contact shifts is generally negligible with respect to pseudocontact one (at least for protons more than three-bonds distant from the metal). This analysis and those conducted on several lanthanide complexes make generally accepted that contact terms might be neglected for ytterbium complexes [2, 14, 16e, 21b, 23, 24, 30].

The choice of Yb(III) is also supported by the small relaxivity of this ion. Indeed, Yb(III) has only one unpaired electron ($^2F_{7/2}$ ground state) and considerably small τ_E, g_J, and J factors (see Table 2.1). Consequently, the NMR spectra of Yb(III) complexes have comparably narrow lines dispersed over a large spectral window (because of the large magnetic susceptibility anisotropy), which leads to resolved multiplets for most resonances of small molecules [24f]. All these aspects contribute to making Yb(III) one of the most suitable metals for using paramagnetic constraints in structural analysis.

2.4. Solution structure determination

2.4.1. The PERSEUS program

The above described relations between the nuclear coordinates and δ^{pc}, T_1, and T_2 ((2.24-25), (2.48-50)) give the theoretical fundament for the use of paramagnetic NMR data as structural constraints in the determination of the solution geometry.

The pseudocontact shift provides geometrical information only once the magnetic anisotropy tensor is known. The principal values of this tensor can be calculated from the crystal field parameters (2.41) [12, 15], but they significantly changes as a function of the ligand cage geometry and can not be correctly determined *a priori*; furthermore, the knowledge of the orientation of the principal axes may not be immediate for rhombic tensors ($D_2 \neq 0$). In principle, also the relaxation times can be calculated from the nuclear coordinates, but in practice the rotational and electronic correlation times rarely are enough accurate to give a satisfactory prediction of the relaxation constants.

The approach used in this thesis is to fit both structure and paramagnetic constants (magnetic anisotropy tensor and relaxation constants) from the experimental data. For each NMR-sensitive nuclei one can collect one δ^{pc} and the T_1 or T_2 data, but they provide only two structural constraints that are not enough to fix the nuclear coordinates (three for each nucleus), the five tensors parameters and the relaxation constants. Thus, without additional information the molecular geometry can not be obtained. In the applications here reported (Chapter 4, 5, and 6) the paramagnetic NMR data are used to adjust only some degrees of freedom of the molecular model and not to determine the complete set of nuclear coordinates. In particular, by keeping constant the distances and the bond angles between the covalently connected nuclei (*i.e.* C−H, C−C distances, H−C−C, C−C−C angles, etc.) and adjusting only the bond conformations (*i.e.* the dihedral angles), the required number of variables strongly reduces, and for small or partially rigid systems these can be sufficient to define a molecular geometry.

The maximum number of structural variables (v) that can be optimised depends on the number of experimental pseudocontact shifts (n_P) and relaxation times (n_T) available, as well as on the tensor symmetry (axial or rhombic). As the $\tilde{\chi}'$ tensor is traceless and symmetric, only five parameters are needed to define it (rhombic case), and in the case of axial symmetry ($\chi'_{xx} = \chi'_{yy}$) they are reduced to three. Calling t the number of parameters needed to define the $\tilde{\chi}'$ tensor and s that one used to fix the relaxation constant,[11] it results that

$$v = n_P + n_T - t - s . \qquad (2.51)$$

In practice, to use a larger number of experimental constraints is preferable, in order to compensate for experimental errors and to obtain more accurate structures.

In this thesis, the structural calculation is obtained through the program PERSEUS (**P**aramagnetic **E**nhanced **R**elaxation and **S**hift for **E**liciting the **U**ltimate **S**tructure) [25]. PERSEUS rearranges the structure of the initial starting model in order to minimize the difference between the calculated and experimental data. The starting model may be a crystal structure as well as the result of a grossly optimised model that approximate the complex structure;[12] if a crystal structure is used the C−H distances are often shorter than their real values [26] and should be corrected [27]. The input model is deformed by rotating around the bonds (ligand conformations), and by moving the metal position; PERSEUS simultaneously

---

[11] We have $s = 1$ if the relaxation times are used, and $s = 0$ if they are not used.

[12] Since PERSEUS optimises the ligand conformations, the starting model may only approximate the real structure. On the contrary, it is important that the bond-lengths and the bond-angles (not optimised by PERSEUS) are set to accurate values (see values reported in Ref. [27]).

fits the $\tilde{\chi}'$ tensor and the relaxation constants. These structural and magnetic parameters are continuously changed until the target function F is minimized

$$F = \frac{\sum_{i=1}^{s} A_i (\delta_i^{pc(calc)} - \delta_i^{pc(exper)})^2}{\sum_{i=1}^{s} (\delta_i^{pc(exper)})^2} + \frac{\sum_{m=1}^{t} B_m (R_m^{par(calc)} - R_m^{par(exper)})^2}{\sum_{m=1}^{t} (R_m^{par(exper)})^2}$$

$$+ \frac{\sum_{l=1}^{u} C_l (r_l - r_l^0)^2}{\sum_{l=1}^{u} (r_l^0)^2} \qquad (2.52)$$

The first two terms of F consider the agreement among the calculated and the experimental pseudocontact shift and relaxation rates.[13] The third term is introduced to force the metal-donors distance r_l to be kept in proximity of the crystallographic bond-length (r_l^0): this term may be used to add another plausible constraint in addition to experimental data. The coefficients A_i, B_m, and C_l adjust the relative weight of each constraint.

The minimization routine in PERSEUS searches the local minimum in the space of the structural and magnetic degrees of freedom around the initial conditions: to be sure of the accuracy of the minimized structure, different initial conditions (tensor values, rotation angles, ...) should be checked in order to arrive as close as possible to the absolute minima.

An index of the agreement between the experimental data and the calculated structure may be represented by the agreement factors $R(\delta^{pc})$ and $R(\rho^{para})$ (Willcott's agreement factors) [28]

$$R(\delta^{pc}) = \sqrt{\frac{\sum_i (\delta_i^{pc(calc)} - \delta_i^{pc(exper)})^2}{\sum_i (\delta_i^{pc(exper)})^2}} \qquad (2.53)$$

$$R(\rho^{para}) = \sqrt{\frac{\sum_i (\rho_i^{para(calc)} - \rho_i^{para(exper)})^2}{\sum_i (\rho_i^{para(exper)})^2}} \qquad (2.54)$$

where the sums run over the i experimental data and ρ^{para} are the paramagnetic relaxation rate R_1^{para} or R_2^{para}.[14]

---

[13] The target function F is general, thus the relaxation rate indices 1 and 2 (relative to the longitudinal and transverse mechanisms) are omitted.

[14] We prefer to change the symbol of the relaxation rates from R^{para} to ρ^{para} to avoid confusion with the symbol R of the Willcott term.

2.4.2. Paramagnetic constraint reliability and dynamics

The paramagnetic constraints offer a unique possibility to investigate the solution structure of ytterbium complexes. In particular the pseudocontact shifts are directly related to the nuclear position through (2.24-25), and they are easily determined from the NMR spectrum. It is remarkable how δ^{pc}'s are very sensitive to small changes in the ligand arrangement: two complexes having slightly different geometries or even two different conformers of the same complex have different NMR spectra.

In Yb(III)-complexes it is often observed that the ligand shifts of a nucleus in two different conformers may be separated by several tens of ppm, which corresponds, at routinely used magnetic fields, to a frequency difference often larger than the conformational exchange rate. In such conditions the conformational dynamics is slow on the NMR time scale, and the resonances of the two species can be distinguished (see Chapter 5 for some examples).

Frequently, processes as conformational equilibria, ligand exchange, chemical equilibria, which are fast, on the NMR time scale, in diamagnetic complexes (as the La^{3+} or Lu^{3+} complexes), are drawn into a slow regime in the analogue Yb(III)-species. In this case, the inspection of EXSY experiments allows one to monitor in detail the conformational dynamics and, through a quantitative analysis of the EXSY integrals, it is possible to estimate the exchange rate of the conformational equilibrium [29, 30].

In slow exchange regime the pseudocontact shifts of the exchanging species are reliable constraints for the structural determination, and can be used to establish the geometry of the single species. At variance, when the exchange is fast, on the NMR time scale, the observed shift is the result of the weighted average of the shifts of the single exchanging species, and does not correspond any more to a reliable geometry.

In fast exchange, information about the rate of the dynamics can be derived from the analysis of the peak linewidth. At variance, structural information can be obtained only once the shifts of the single species are extrapolated. A common example is the case of a ligand in fast exchange between a free form and a form bound to the metal complex: the ligand resonances are averaged to a shift that is weighted on the relative amounts of free and bound forms. With an NMR titration of the ligand, the relative amounts of the two species changes and the ligand shifts move towards the resonances of the more abundant species: thus, the analysis of the titration curves allows one to extrapolate the shifts of the pure bound form. In turn, the shifts of the bound form can be properly used for structural determination (see examples in Chapter 6).

Otherwise, fast or intermediate exchanges can often be drawn into a slow exchange by lowering the temperature. This procedure is especially effective in Yb(III)-complexes because lowering the temperature not only reduces the exchange rate, but it also increases the anisotropy of the magnetic susceptibility (following the T^{-2} trend), and consequently the frequency difference among the resonances of the exchanging species.

In conclusion, the reliability of the paramagnetic constraints depends on the presence of dynamics that can affect them. In particular, is important to exclude the presence of fast exchange before the use of paramagnetic data (especially δ^{pc}) in a structural calculation. An important alarm bell is the inspection of the peak linewidth: peaks larger than what is expected on the bases of the pure paramagnetic relaxation (see (2.48-50)), strongly suggest the presence of dynamics or chemical equilibria, and it requires a specific discussion.

In the slow exchange conditions as well as in the absence of chemical or conformational exchanges, the solution structure obtained from the fitting of the paramagnetic constraints are the best solution model compatible with the experimental data: the examples reported in Chapter 4-6 show structures with errors around 0.1-0.2 Å. As the pseudocontact shifts and the relaxation rates are larger in proximity of the metal, the precision of the calculated structure becomes better near the metal. In spite of this, nuclei that are three or less bonds distant from the metal may have significant contact shift contributions, which should be discussed or considered in the optimisation.

Finally, even in absence of chemical or conformational equilibria, a molecule in solution undergoes librations or small motions that are generally much faster than the conformational exchange, and are not resolved by the paramagnetic interaction. This kind of mobility probably represent the very limit in the modelling of the solution structure; anyway, in most the cases the displacement range in the nuclear position due to these motions is limited, so that the error reflected in the pseudocontact shift (or in the relaxation times) is comparably small, allowing one to fit accurate geometries.

References

[1] (a) BERTINI, I.; LUCHINAT, C., *Coord. Chem. Rev.* **1996**, *150*. (b) BERTINI, I.; LUCHINAT, C.; PARIGI, G., *Solution NMR of Paramagnetic Molecules*, **2001**, Elsevier: Amsterdam. (c) BERTINI, I.; LUCHINAT, C.; PARIGI, G., *Prog. Nucl. Magn. Reson. Spec.* **2002**, *40*, 249–273.

[2] PETERS, J. A.; HUSKENS, J.; RABER, D. J., *Prog. Nucl. Magn. Reson. Spec.* **1996**, *28*, 283–350.

[3] PIGUET, C.; GERALDES, C. F. G. C., in *Handbook on the Physics and Chemistry of Rare Earths*; Gschneidner, K. A. Jr.; Bünzli, J.-C. G.; Pecharsky, V. K., Eds., **2003**, Elsevier: Amsterdam. Vol. 33, p. 353–463.

[4] GOLDING, R. M.; PASCUAL, R. O.; MCGARVEY, B. R., *J. Magn. Reson.* **1982**, *46*, 30–42.

[5] SOLOMON, I., *Phys. Rev.* **1955**, *99*, 559.

[6] ABRAGAM, A., *The Principles of Nuclear Magnetism*, **1961**, Oxford University Press: Oxford.

[7] KOENIG, S. H., *J. Magn. Reson.* **1982**, *47*, 441–453.

[8] GUÉRON, M., *J. Magn. Reson.* **1975**, *19*, 58–66.

[9] VEGA, A. J.; FIAT, D., *Mol. Phys.* **1976**, *31*, 347–355.

[10] BLOEMBERGEN, N., *J. Chem. Phys.* **1957**, *27*, 575.

[11] PETERS, J. A.; NIEUWENHUIZEN, M. S.; RABER, D. J., *J. Magn. Reson.* **1985**, *65*, 417.

[12] BLEANEY, B. „*J. Magn. Reson.* **1972**, *8*, 91–100.

[13] MCGARVEY, B. R., *J. Magn. Reson.* **1979**, *33*, 445–455.

[14] KEMPLE, M. D.; RAY, B. D.; LIPKOWITZ, K. B.; PRENDERGAST, F. G.; RAO, B. D. N., *J. Am. Chem. Soc.* **1988**, *110*, 8275–8287.

[15] (a) MIRONOV, V. S.; GALAYAMETDINOV, Y. G.; CEULEMANS, A.; GOERLLER-WALRAND, C.; BINNEMANS, K., *J. Chem. Phys.* **2002**, *116*, 4673–4685. (b) MIRONOV, V. S.; GALAYAMETDINOV, Y. G.; CEULEMANS, A.; GOERLLER-WALRAND, C.; BINNEMANS, K., *Chem. Phys. Lett.* **2001**, *345*, 132–140.

[16] (a) ALSAADI, B. M.; ROSSOTTI, F. J. C.; WILLIAMS, R. J. P., *J. Chem. Soc., Dalton. Trans.* **1980**, 2151–2154. (b) MISRA, S. K.; ORHUM, U., *Solid State Commun.* **1987**, *63*, 867. (c) MALHOTRA, V. M.; BUCKMASTER, H. A.; DIXON, J. M., *J. Phys. C: Solid State Phys.* **1980**, *13*, 3921. (d) BERTINI, I.; CAPOZZI, F.; LUCHINAT, C.; NICASTRO, G.; XIA Z., *J. Phys. Chem.* **1993**, *101*, 198. (e) AIME, S.; BOTTA, M.; ERMONDI G., *Inorg. Chem.* **1992**, *31*, 4291.

[17] ORBACH, R., *Proc. R. Soc. London* **1961**, *SER. A264*, 458.

[18] (a) VISHNEVSKAYA, G. P., KOZYREV, D. M., *J. Struct. Chem.* **1966**, 7, 20. (b) HUDSON, A.; LEWIS, J. W. E., *Trans. Faraday Soc.* **1970**, *66*, 1297. (c) KOENIG, S. H., *Magn. Reson. Med.* **1991**, *22*, 183. (d) HERNANDEZ, G.; TWEEDLE, M.; BRYANT R. G., *Inorg. Chem.* **1990**, *29*, 5110. (e) AIME, S.; BARBERO, L.; BOTTA M., *Magn. Res. Imaging* **1991**, *9*, 843.

[19] (a) REUBEN, J.; ELGAVISH, G. A., *J. Magn. Reson.* **1980**, *39*, 421–430. (b) REUBEN, J., *J. Magn. Reson.* **1982**, *50*, 233–236. (c) PLATAS, C.; AVECILLA, F.; DE BLAS, A.; GERALDES, C. F. G. C. ; RODRÍGUEZ-BLAS, T.; ADAMS, H.; MAHIA, J., *Inorg. Chem.* **1999**, *38*, 3190–3199. (d) RIGAULT, S.; PIGUET, C., *J. Am. Chem. Soc.* **2000**, *122*, 9304–9305. (e) GERALDES, C. F. G. C.; ZHANG, S.; SHERRY, A. D., *Inorg. Chim. Acta* **2004**, *357*, 381.

[20] (a) OUALI, N.; RIVERA, J.-P.; CHAPON, D.; DELANGLE, P.; PIGUET, C., *Inorg. Chem.* **2004**, *43*, 1517–1529. (b) TERAZZI, E.; RIVERA, J.-P.; OUALI, N.; PIGUET, C., *Magn. Reson. Chem.* **2006**, *44*, 539–552.

[21] (a) DESREUX, J. F.; REILLEY, C. N., *J. Am. Chem. Soc.* **1976**, *98*, 2105. (b) REILLEY, C. N.; GOOD, B. W.; ALLENDOERFER, R. D., *Anal. Chem.* **1976**, *48*, 1446.

[22] PETERS, J. A., *J. Magn. Reson.* **1986**, *68*, 240–251.

[23] FORSBERG, J. H.; DELANEY, R. M.; ZHAO, Q.; HARAKAS, G.; CHANDRAN, R., *Inorg. Chem.* **1995**; *34*, 3705–3715.

[24] (a) REILLEY, C. N.; GOOD, B. W.; DESREUX, J. F., *Anal. Chem.* **1975**, *47*, 2110. (b) DESREUX, J. F., *Inorg. Chem.* **1980**, *19*, 1319–1324. (c) STAINER, M. V. R.; TAKATS, J., *J. Am. Chem. Soc.* **1983**, *105*, 410. (d) BRITTAIN, H. G.; DESREUX, J. F., *Inorg. Chem.* **1984**, *23*, 4459–4466. (e) SHELLING, J. G.; BJORNSON, M. E.; HODGES, R. S.; TENEJA, A. K.; SYKES, B. D., *J. Magn. Reson.* **1984**, *57*, 99. (f) DI BARI, L.; LELLI, M.; PINACUDA, G.; PESCITELLI, G.; MARCHETTI, F.; SALVADORI, P., *J. Am. Chem. Soc.* **2003**, *125*, 5549–5558.

[25] DI BARI, L.; PINTACUDA, G.; RIPOLI, S.; SALVADORI, P., *Magn. Reson. Chem.* **2002**, *40*, 396–405.

[26] The crystallographic C–H distances are around 0.93 Å, shorter than the effective bond length [CHURCHILL, M. R., *Inorg. Chem.* **1973**, *12*, 1213–1214.]. The values determined with more accurate method are reported in Ref. 27 and in HENRY, B. R., *Acc. Chem. Res.* **1987**, *20*, 429–435.; BARTELL, L. S.; ROTH, E. A.; HOLLOWELL, C. D.; KUCHITSU, K.; YOUNG, J. E. J., *J. Chem. Phys.* **1965**, *42*, 2683–2686; and references therein.

[27] MARCH, J., *Advanced Organic Chemistry Reaction Mechanism and Structures*, **1992**, 4th ed.; Wiley Interscience: New York.

[28] (a) WILLCOTT, M. R.; LENKINSKI, R. E.; DAVIS, R. E., *J. Am. Chem. Soc.* **1972**, *94*, 1742–1744. (b) DAVIS, R. E.; WILLCOTT, M. R., *J. Am. Chem. Soc.* **1972**, *94*, 1744–1745.

[29] PERRIN, C. L.; GIPE, R. K., *J. Am. Chem. Soc.* **1984**, *106*, 4036–4038.

[30] DI BARI, L.; PINTACUDA, G.; SALVADORI, P., *Eur. J. Inorg. Chem.* **2000**, 75–82.

Chapter 3
Near-IR circular dichroism of Yb(III) complexes

In Yb(III) complexes it is possible to observe the CD related to the $f \rightarrow f$ electronic transitions of the metal. These fall around 980 nm in the near-IR range (NIR), and are scarcely influenced by the UV-Vis transitions typical of organic ligands, so the NIR CD can be detected even in molecules having strong UV-Vis absorptions. While the UV-Vis CD provides information on the chiral arrangement of the *ligand* [1], Yb(III) NIR CD offers a complementary point of view giving information about the chiral distortion of the coordination polyhedron *around the metal*.

In the present chapter we report some theoretical elements on the optical transitions in Yb(III) ions (Section 3.1) and how these may be used to provide precious information about the metal coordination (Section 3.2), in order to introduce this technique largely used in Chapter 4, 5, and 6.

3.1. Some elements of theory

Circular Dichroism is the difference between the extinction molar coefficients measured using left (ε_L) and right (ε_R) circularly polarized light [2]

$$\Delta\varepsilon = \varepsilon_L - \varepsilon_R \tag{3.1}$$

where the absorption spectrum measure the average value

$$\varepsilon = \frac{\varepsilon_L + \varepsilon_R}{2} . \tag{3.2}$$

In absence of external fields, CD arises only in the presence of chiral *non-racemic* compounds and it has opposite sign for the two enantiomers.

Considering a single transition between the ground state $|A_0\rangle$ and the excited one $|A_a\rangle$, the intensities of the absorption and of the corresponding CD bands (also called Cotton effect) depend on the oscillator strength

(D_{0a}) and the rotational strength (R_{0a}), respectively [3]

$$D_{0a} = \frac{3hc10^3 \ln 10}{8\pi^3 N_A} \int \frac{\xi_0 \varepsilon}{\tilde{v}} d\tilde{v} \tag{3.3}$$

$$R_{0a} = \frac{3hc10^3 \ln 10}{32\pi^3 N_A} \int \frac{\xi_0 \Delta\varepsilon}{\tilde{v}} d\tilde{v} \tag{3.4}$$

where h is the Planck constant, c is the vacuum light speed, N_A is the Avogadro number, and $\tilde{v}$ is the light wave number, ξ_0 is the degeneracy of the starting level, the numerical factor 10^3 is introduced to accommodate the extinction molar coefficients in the commonly used $M^{-1}cm^{-1}$ units. Both the oscillator strength D_{0a} and the rotational strength R_{0a} are related to the electric ($\mathbf{P}_{0a} = \langle A_0| \hat{\boldsymbol{\mu}} |A_a\rangle$) and magnetic transition dipoles ($\mathbf{M}_{0a} = \langle A_0| \hat{\mathbf{m}} |A_a\rangle$)

$$D_{0a} = |\mathbf{P}_{0a}|^2 + |\mathbf{M}_{0a}|^2 \tag{3.5}$$
$$R_{0a} = \text{Im}(\mathbf{P}_{0a} \cdot \mathbf{M}_{0a}) \tag{3.6}$$

and the rotational strength corresponds to the imaginary part of the scalar product of the electric and magnetic transition dipole. The intensity of the Cotton effect with respect to the absorption band is quantified by the dissymmetry factor g

$$g = \frac{\Delta\varepsilon}{\varepsilon} \tag{3.7}$$

which is in turn proportional to the R_{0a}/D_{0a} ratio. If $|\mathbf{P}_{0a}|^2 >> |\mathbf{M}_{0a}|^2$, we can write [4]

$$g_{0a} = \frac{4R_{0a}}{D_{0a}} = \frac{4\text{Im}(\mathbf{P}_{0a} \cdot \mathbf{M}_{0a})}{|\mathbf{P}_{0a}|^2 + |\mathbf{M}_{0a}|^2} \approx \frac{4\text{Im}(\mathbf{P}_{0a} \cdot \mathbf{M}_{0a})}{\text{Re}(\mathbf{P}_{0a} \cdot \mathbf{P}_{0a}^*)} \approx \frac{4|\mathbf{M}_{0a}|}{|\mathbf{P}_{0a}|}\cos(\eta) \tag{3.8}$$

where η is the angle between the transition dipole vectors $\mathbf{P}_{0a}$ and $\mathbf{M}_{0a}$. As a consequence, in order to have an appreciable CD, the observed transition must be magnetically allowed.

3.1.1. Spectroscopic states in Ln(III) ions

The optical properties of most of the Ln(III) ions are explained by terms of **$4f \rightarrow 4f$ intraconfigurational radiative transitions**.

The electronic Hamiltonian for the free lanthanide ion, limited to the $4f$ shell, is described by four terms, whose H_{4f}^C and H_{4f}^{SO} account for the electron-electron coulombic repulsion and the spin-orbit coupling, respectively

$$H_{4f}^{\text{free}} = -\frac{\hbar^2}{2m} \sum_{i=1}^{N} \nabla_i^2 - \sum_{i=1}^{N} \frac{Z^* e^2}{r_i} + \underbrace{\sum_{i<j}^{N} \frac{e^2}{r_{ij}}}_{H_{4f}^{C}} + \underbrace{\sum_{i=1}^{N} \zeta(r_i) \mathbf{s}_i \cdot \mathbf{l}_i}_{H_{4f}^{SO}} \quad (3.9)$$

where $\hbar = h/2\pi$, m and e are the electron mass and charge, respectively, Z^* is the shielded nuclear charge, r_i and r_{ij} are the electron-nucleus and the electron-electron distances, $\zeta(r_i)$ is the spin-orbit function, and $\mathbf{s}_i$ and $\mathbf{l}_i$ are the spin and orbital angular moment operators for the single electron, and all the sums run over the N electrons of the $4f$ orbitals. Depending on the relative importance of the terms H_{4f}^{C} and H_{4f}^{SO} two approximations are possible:

- for $H_{4f}^{C} > H_{4f}^{SO}$ we are in the **Russell-Saunders coupling**, and the spin-orbit coupling is considered as a small perturbation on the electric level structure determined by diagonalizing H_{4f}^{C};
- for $H_{4f}^{C} < H_{4f}^{SO}$ occurs the so called *j-j* **coupling**, in which the spin and angular moment are coupled in the J quantum number ($|J| = |L \pm S|$), and the coulombic interaction perturbs the J scheme.

The case of lanthanides is an intermediate situation (**intermediate coupling**), in which the Russell-Saunders coupling can be adopted only as a first approximation, and successively one must evaluate the effect of the spin-orbit coupling that generates a separation of the spectroscopic $^{(2S+1)}L$ states, on the basis of the different quantum numbers J.

The combination of these interactions is responsible for the complex system of $4f$ levels observed for the Ln(III) ions, which is depicted in Figure 3.1 [5]. In some cases (as for Eu^{3+}) there is a large number of states, which gives rise to very complicated electronic spectra. In Ln^{3+} complexes, the $4f$ orbitals have very scarce overlap with the ligand orbitals (see Chapter 1), as a consequence, the presence of the ligand introduces only a weak interaction with the $4f$ orbitals, which can be accounted for on the bases of the crystal-field theory. In the crystal-field theory, the electronic Hamiltonian of the complex (H_{4f}^{TOT}) is calculated from the free-ion Hamiltonian by introducing the perturbation due of the crystal-field term H_{4f}^{CF}

$$H_{4f}^{\text{TOT}} = H_{4f}^{\text{free}} + H_{4f}^{CF} \quad (3.10)$$

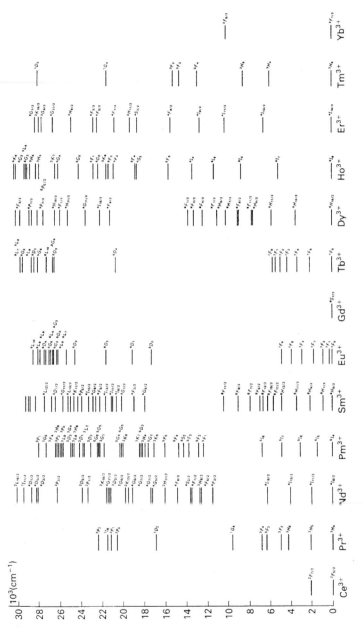

Figure 3.1. Energy of the electronic levels in the Ln^{3+} free ions.

The crystal-field Hamiltonian is in turn decomposed into the sum

$$H_{4f}^{CF} = \sum_{kq} B_q^k \mathbf{C}_q^k \tag{3.11}$$

where $\mathbf{C}_q^k$ are the intraconfigurational spherical tensor operators of rank k and order q, and B_q^k are the corresponding crystal field parameters.

The action of the crystal field removes the degeneracy of the J-levels, splitting the free-ion electronic levels into sublevels [6]. The direct consequence is that the electronic transitions also split into a series of bands. The intensity of the crystal-field energy split depends on the crystal field parameters, and is a function of both the donor nature and the complex geometry.

Figure 3.2 schematically shows the typical absorption region for each Ln(III) ions [7]: the spectra cover all the UV-Vis-NIR and IR regions. For complexes having absorption in the Visible region, these transitions are also responsible of the complex color.

As many of these transitions are electric dipole forbidden ($\Delta L = 0$) generally the absorption spectra are weak, only Tb^{3+} and Ce^{3+} have molar extinction coefficients ε as higher as 10^2. Because the $4f$ orbitals are well shielded from ligands, vibrational couplings are limited and the absorption bands are comparably sharp.

3.1.2. f–f transitions in Yb(III)

The spectroscopic states of Yb^{3+} free ion are composed by only two terms: $^2F_{7/2}$ for the ground state, and $^2F_{5/2}$ for the excited one, and are separated by an energy gap about 980 nm (NIR range).

The crystal-field of the ligand splits the ground and the excited state in two patterns of four and three sublevels, respectively. The energy separation among the level in the ground-state pattern is of the order of kT (207 cm^{-1} at 298 K), so all the four sublevels of the $^2F_{7/2}$ configuration are considerably populated at room temperature (Figure 3.3).

This allows up to 12 possible transitions in absorption and CD spectra whose center of gravity is around 980 nm. All the transitions are electric dipole forbidden ($\Delta L = 0$) and many are magnetic dipole allowed ($\Delta J = 1$): this reflects in weak absorptions (due probably with some coupling with $5d$ levels) [7] and comparably high rotational strength (3.6), with dissymmetry factors g that can be $g > 0.25$ [9].

The high g-values and the absence of interferences from the solvent absorption bands make the CD spectra easy to acquire and sensitive, com-

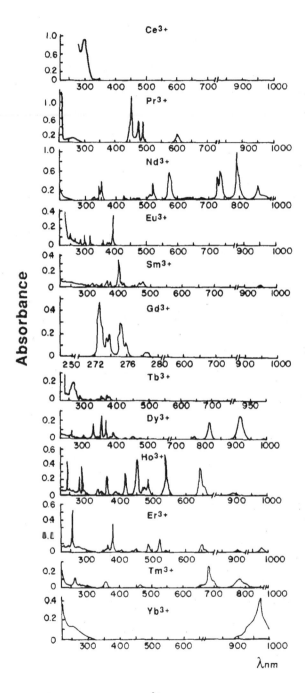

Figure 3.2. Absorptions spectra of Ln^{3+} ions in solution down to 1000 nm.

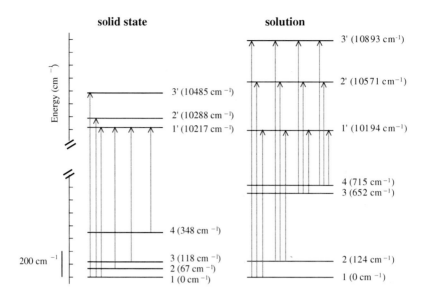

Figure 3.3. Diagram of the energy levels of Yb^{+3} determined both at the solid state ($Na_3[Yb(dpa)_3] \cdot 13H_2O$) [8] and in solution ($[Yb((R)\text{-}DOTMA)]$) [9]. The arrows show the transitions assigned in the corresponding NIR spectrum.

pared to the corresponding absorption spectra.[1] Furthermore, in the CD spectra, it is often possible to resolve overlapping bands when the sign of their Cotton effects are opposite. This is evident looking at the CD and absorption spectra of the macrocyclic complex $[Yb((R)\text{-}DOTMA)]$[2] (Figure 3.4). The enhanced resolution of its CD spectrum allowed the determination of all the 12 transition bands and the assignment of the electronic levels (Figure 3.3) [9].

On lowering the temperature two distinct effects are observed in the spectra: (i) the spectral lines are narrowed, (ii) the population of the lower-lying sublevel of the ground state increases, with a consequent increase of the CD intensity of transitions originating from this level. As a consequence, by comparing the intensity of the CD-bands in spectra recorded at room temperature (298 K) and at 193 K it is possible to iden-

---

[1] In the NIR region often fall vibrational overtones of the common solvents. These absorptions are more intense of the Yb(III) bands, and the not perfect subtraction of the solvent bands introduce artefact and reduces the quality of the spectrum. On the contrary, the CD is sensitive only to the chiral compounds, and the solvent does not give any bands, so the transitions of the complex are easily recognized.

[2] (*R*)-DOTMA is (1*R*,4*R*,7*R*,10*R*)-$\alpha,\alpha',\alpha'',\alpha'''$-tetramethyl-1,4,7,10-tetraazacyclododecane-1,4, 7,10-tetraacetic acid.

tify the transitions starting from the ground state, that is the first step in the assignment of the whole spectral pattern (see also Section 4.5.2) [9].

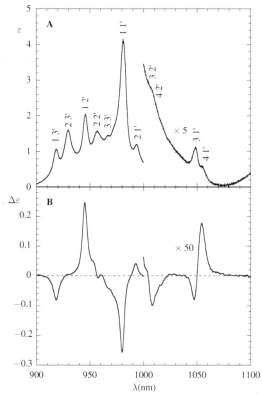

Figure 3.4. NIR absorption **A** and CD spectra **B** of [Yb((*R*)-DOTMA)], (*c* = 0.16 M, pathlength 1 cm). The second part of each spectrum is expanded by a factor of 5 **A** and 50 **B**.

3.2.   Yb NIR CD for the structural determination

The intensity, sign and frequencies of the NIR CD Cotton effects are related to the crystal field coefficients (3.11), which reflect the nature and the arrangement of the coordination polyhedron around the metal. Ytterbium complexes that have the similar coordination polyhedron have also similar NIR CD spectra. Examples of this are found comparing the NIR CD spectra of macrocyclic [Yb(DOTA-like)] complexes. Indeed, in these molecules the coordination polyhedron can assume a square antiprismatic (*SA*) or a twisted-square antiprismatic (*TSA*) geometry (see Section 5.1 for further details). It is demonstrated that complexes as [Yb((*R*)-DOTAMPh)], [Yb((*S*)-DOTAMNp)], and [Yb(*S*-(*RRRR*)-

NO$_2$BnDOTMA)] have a *SA* arrangement in solution [10, 11], while [Yb((R)-DOTMA)], and [Yb(S-($SSSS$)-NO$_2$BnDOTMA)] are essentially *TSA* [11].[3] The NIR CD spectra of these systems show that the spectra of the *SA* forms are each other closely related, as well as those of the *TSA* forms: but the *TSA* and the *SA* NIR CD spectra are largely different, demonstrating the relation between geometry and NIR CD spectrum. This is also true for other Yb-systems as the heterobimetallic complexes Na$_3$ [Yb((S)-BINOL)$_3$] and K$_3$[Yb((S)-BINOL)$_3$] [12] (further details are in Section 4.5).

All these examples indicate that NIR CD is essentially sensitive to the arrangement of the first coordination sphere, and is scarcely affected by the structural changes far from the metal that let unchanged geometry and charge of the coordination polyhedron.

On the other side, if the coordination sphere is perturbed, *e.g.* introducing an axial ligand in the macrocyclic complexes, the CD immediately reflects this change. The gradual addition of an axial ligand (as DMSO) to the [Yb((R)-DOTAMPh)] and [Yb((S)-DOTAMNp)] complexes produce a drastic change in the band intensity around 980 nm (Figure 3.5), and this was demonstrated to occur in concomitance with the axial coordination of the DMSO [10, 13].

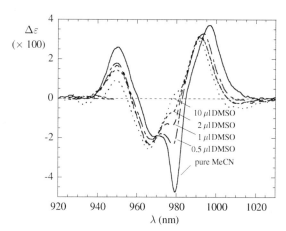

Figure 3.5. NIR CD spectra of the titration of [Yb((R)-DOTAMPh)] in CH$_3$CN (c = 20 mM, pathlength = 1 cm) with DMSO (addition from 0.5 to 10 μL in a 0.5 mL of solution of the complex).

---

[3] **DOTAMPh** is 1,4,7,10-tetrakis[(R)-1-(phenyl)ethylcarbamoilmethyl]-1,4,7,10-tetraazacyclodo-decane; **DOTAMNp** is 1,4,7,10-tetrakis[(S)-1-(1-naphthyl)ethylcarbamoilmethyl]-1,4,7,10-tetra-azacyclododecane; **S-($RRRR$)-NO$_2$BnDOTMA** is (1R,4R,7R,10R)-$\alpha,\alpha',\alpha'',\alpha'''$-tetramethyl-[($S$)-2-(nitrobenzyl)]-1,4,7,10-tetraazacyclododecane-1,4,7,10-tetraacetic acid; **S-($SSSS$)-NO$_2$Bn DOTMA** is (1S,4S,7S,10S)-$\alpha,\alpha',\alpha'',\alpha'''$-tetramethyl-[($S$)-2-(nitrobenzyl)]-1,4,7,10-tetraazacy-clododecane-1,4,7,10-tetraacetic acid.

In Chapter 5, we shall treat in further examples how the analysis of the analogies/differences among the NIR CD spectra of chiral [Yb(DOTA-like)] complexes makes possible the assignment of the arrangement of coordination polyhedron, providing an essential piece of information for the structural determination [14].

References

[1] (a) HARADA N., NAKANISHI K., *Circular Dichroic Spectroscopy-Exciton Coupling in Organic Stereochemistry*, **1983**, University Science Books: Mill Valley, CA. (b) NAKANISHI, K.; BEROVA, N.; WOODY, R. W., *Circular Dichroism: principles and applications*, **2000**, 2nd ed.; Wiley-VCH: New York.

[2] ELIEL, E. L.; WILEN, S. H., *Stereochemistry of Organic Compounds*, **1994**, Wiley & Sons, Inc.: New York.

[3] MASON, S. F., *Molecular Optical Activity & the Chiral Discrimination*, **1982**, Cambridge University Press: Cambridge.

[4] RICHARDSON, F. S.; FAULKNER, T. R., *J. Chem. Phys.* **1982**, *76*, 1595–1606.

[5] HÜFNER, S., *Optical spectra of transparent Rare Earth Compounds*, **1978**, Academic Press: New York.

[6] RICHARDSON, F. S., *Inorg. Chem.* **1980**, *19*, 2806–2812.

[7] EVANS C. H., *Biochemistry of the Lanthanides*, **1990**, Plenum Press: New York.

[8] REINHARD, C.; GÜDEL, H., *Inorg. Chem.* **2002**, *41*, 1048–1055.

[9] DI BARI, L.; PINTACUDA, G.; SALVADORI, P., *J. Am. Chem. Soc.* **2000**, *122*, 5557–5562.

[10] DI BARI, L.; PINTACUDA, G.; SALVADORI, P.; DICKINS, R. S.; PARKER, D., *J. Am. Chem. Soc.* **2000**, *122*, 9257–9264.

[11] DI BARI, L.; PESCITELLI, G.; SHERRY, A. D.; WOODS, M., *Inorg. Chem.* **2005**, *44*, 8391–8398.

[12] DI BARI, L.; LELLI, M.; PINTACUDA, G.; PESCITELLI, G.; MARCHETTI, F.; SALVADORI, P., *J. Am. Chem. Soc.* **2003**, *125*, 5549–5558.

[13] DICKINS, R. S.; PARKER, D.; BRUCE, J. I.; TOZER, D. J., *Dalton Trans.* **2003**, 1264 1271.

[14] LELLI, M.; PINTACUDA, G.; CUZZOLA, A.; DI BARI, L., *Chirality*, **2005**, *17*, 201–211.

PART II

Chapter 4
Structural study of Yb-heterobimetallic catalysts

Heterobimetallic lanthanoid catalysts are a family of Ln(III) complexes of general formula $M_3[Ln(BINOL)_3]$, (BINOL is enantiopure 1,1'-bis(2-naphtholate),[1] and the alkaline metals are M = Li, Na, K, Figure 4.1). The first complex synthesised was $Li_3[La(BINOL)_3]$, introduced by Shibasaki *et al.* in 1992 as stereoselective catalyst for the nitroaldolic reaction [1].

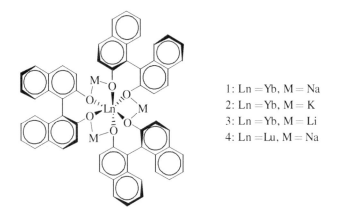

1: Ln = Yb, M = Na
2: Ln = Yb, M = K
3: Ln = Yb, M = Li
4: Ln = Lu, M = Na

Figure 4.1. Formula of the heterobimetallic catalysts used.

From that date on, Shibasaki and co-workers have found that these systems are effective stereoselective precursors for many other reactions such as the aldol-type condensations, the Michael reaction, imines and aldehydes hydrophosphonylation, the Diels-Alder reaction, the epoxidation of enones, etc. (Figure 4.2) [2, 3]. Independently, Greeves *et al.* studied the application of such systems to the stereocontrolled addition of alkyl-lithium reagents to aldehydes [3, 4].

---

[1] BINOL-H_2 was used to indicate the enantiopure binaphthol.

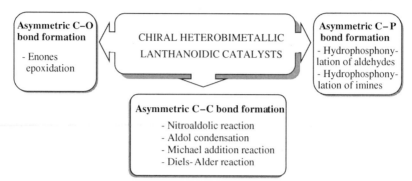

Figure 4.2. Some important applications of heterobimetallic lanthanoid complexes in asymmetric catalysis.

The high performances of these catalysts [2], and the application of such systems to important and diverse reactions [2, 5], justify the interest toward the structural study and the catalytic mechanism of these systems.

In this chapter, we focused our attention on the ytterbium heterobimetallic catalysts $Na_3[Yb((S)\text{-BINOL})_3]$ (**1**), $K_3[Yb((S)\text{-BINOL})_3]$ (**2**), $Li_3[Yb((S)\text{-BINOL})_3]$ (**3**), exploiting a solution structural study through NMR and NIR CD spectroscopies. The determined solution structures were compared to the crystallographic structures discussing their differences. The solution investigation was also extended to a study of the ligand lability, which introduced new elements about the behaviour of such systems that constitute the bases for a further investigation of the catalytic mechanism.

4.1. Catalytic activity and reaction mechanism

4.1.1. Stereoselective nitroadolic reaction

The stereoselective nitroadolic condensation (Henry's reaction, Figure 4.3) was the first example of heterobimetallic lanthanoidic catalysis. Lanthanide alkoxides, such as $La_3(OtBu)_9$, were well known to promote Henry's reaction [6], but only when the alkoxides were replaced by chiral binaphtholate in $Li_3[La((S)\text{-BINOL})_3]$ the reaction proceeded with high yield and e.e. (Figure 4.3) [1].

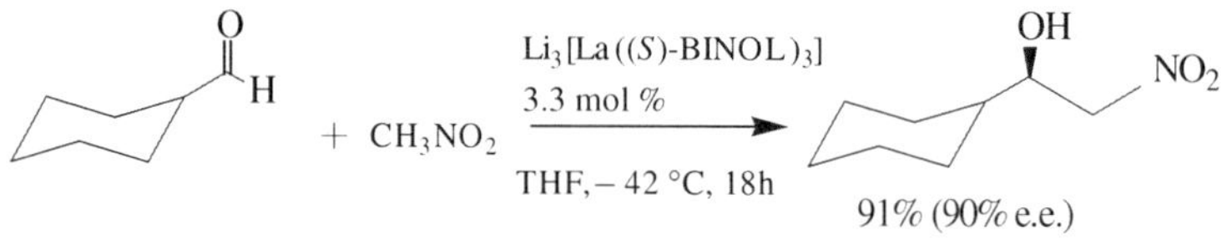

Figure 4.3. Enantioselective nitroaldol reaction catalysed by $Li_3[La((S)\text{-BINOL})_3]$.

The versatility of both hydroxyl and nitro groups, makes this reaction of great interest in the synthesis of amino-alcohols for pharmacological and biological applications [5].

The enantioselectivity depends on the lanthanide ion and the alkali metal that compose the catalyst: the lithium complexes induce high e.e., whereas the sodium catalysts produce almost racemates [7]. Figure 4.4 reports the trend of the e.e. obtained through catalysts of different lanthanide ions: the best results are generally obtained with metals of the first part of the series, but there are considerable differences depending on the substrates used. In some cases, an additional improvement of the catalytic efficiency can be achieved using binaphtholate substituted in position 6 and 6' [8].

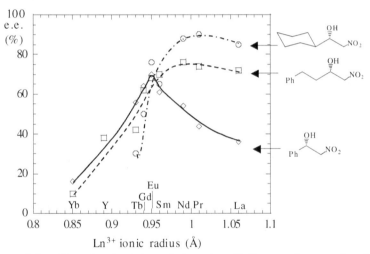

Figure 4.4. Enantioselectivity of the nitroaldolic reaction catalysed by $Li_3[Ln((R)\text{-BINOL})_3]$ as a function of the Ln^{3+} ionic radius and the substrates used [2b].

In order to rationalize these results the mechanism reported in Figure 4.5 was proposed [9]. The coordination of nitromethane (first step) involves the interaction with both the lanthanide and the alkali metal ion: the combined effects of the acidic character of the Ln^{3+} ions and the basicity of the binapholate promote the proton extraction and the nitronate enolization (adduct **II**). The coordination of the aldehyde (**III**) activates the carbonyl for nitronate insertion (**IV**). The cycle concludes with the proton exchange and the release of the products that regenerate the catalyst (**I**).

The first step, with the formation of the nitronate, is supposed to be the rate-determining step [9], and it is directly influenced by the nature the lanthanide and the alkali metal used. The lithium complexes are probably

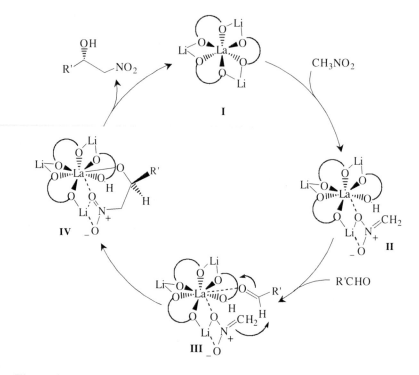

Figure 4.5. Catalytic cycle proposed for the nitroaldolic addition.

the best systems because they are basic enough to promote enolization and stabilize the nitronate.[2]

The chiral discrimination depends on the relative orientation between the nitronate and the aldehyde, which is induced by the chiral ligand environment in the adduct **III**. Models based on the crystal structure of Na$_3$[La((S)-BINOL)$_3$] support the stereoselective insertion of the nitronate (**III**→**IV**), but experimental evidences of the intermediates **III-IV** are still lacking [9]. It was suggested [7] that little structural differences among the complexes of different Ln^{3+} ions are probably the cause of the different enantioselectivity found for each substrate and might justify the trends observed in Figure 4.4.

---

[2] A "second generation" of heterobimetallic catalysts was built enriching the catalysts Li$_3$[La(BINOL)$_3$] with 1 equivalent of LiOH. This new catalyst showed an increased activity: with 1.0 mol% of Li$_3$[La((S)-BINOL)$_3$]·LiOH one obtains results comparable to those reported using 3.3 mol% of Li$_3$[La(BINOL)$_3$] ([2]). The enhanced basicity of the second-generation catalyst is supposed to promote the formation of the nitronate species. FAB MS spectrometry evidenced the formation of an adduct between the nitronate and a fragment of Li$_3$[La((S)-BINOL)$_3$]·LiOH ([9]).

4.1.2. Michael addition reaction

Heterobimetallic catalysts are effective also in the asymmetric aldol condensation and in the Michael reaction (Figure 4.6) [10, 11].

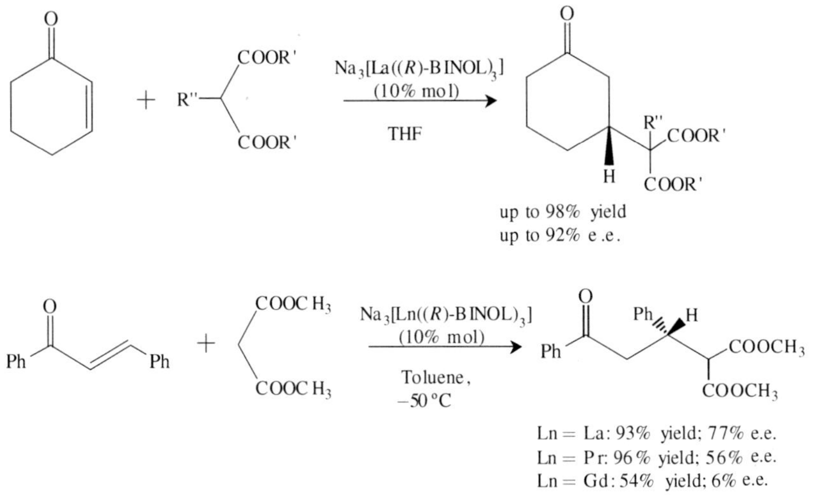

Figure 4.6. Examples of enantioselective Michael addition.

The sodium complexes are generally the best catalysts for this reaction (often with e.e. > 90%), at variance of the lithium systems that give nearly racemic products [11].

Generally, the Ln^{3+} ions of the beginning of the series give higher e.e. (especially $Na_3[La(BINOL)_3]$ and $Na_3[Pr(BINOL)_3]$), with significant differences depending on the substrates [11]. The solvent plays an important role: pure THF for some substrates must be replaced by more apolar mixture as THF/toluene, toluene or CH_2Cl_2 [2, 11].

The proposed catalytic cycle is similar to the one seen for the nitroaldol reaction (Figure 4.7). The ambivalent basic and Lewis acid character of these catalysts promotes the enolization of the malonate and the activation of the enone, respectively (**II**). Both the substrates remain bound to the catalyst that induces the stereocontrolled insertion of the malonate (**III**). The malonate remains anchored to the complex through the sodium ion and transfers a proton to the binaphtholate (**II**). Probably, in THF the malonate may dissociate from the catalyst attacking the enone directly "from the solution" and without stereocontrol; for this reason, in less polar solvents as CH_2Cl_2 or toluene, where this dissociation is prevented, higher enantiomeric excess is obtained. Last step is the proton exchange that regenerates the free catalyst (**I**).

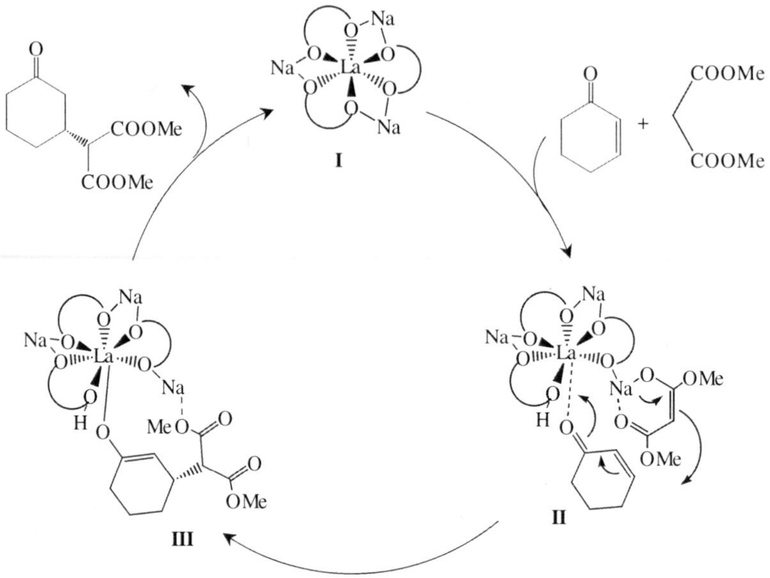

Figure 4.7. Catalytic cycle proposed for the Michael addition.

Shibasaki *et al.* reported NMR evidences of the interaction of the enone with the catalyst [11]; furthermore, computer simulations were used to support the malonate insertion on the ciclohexanone (**III**) [11].

4.1.3. Hydrophosphonylation of imines

Heterobimetallic catalysts are active in the enantioselective hydrophos-phonylation of the imines [12] and aldehydes [13, 14, 15]. This reaction leads to the formation of new C−P bond through the addition of alkyl phosphonates to the C=N and C=O double bond (Figure 4.8).

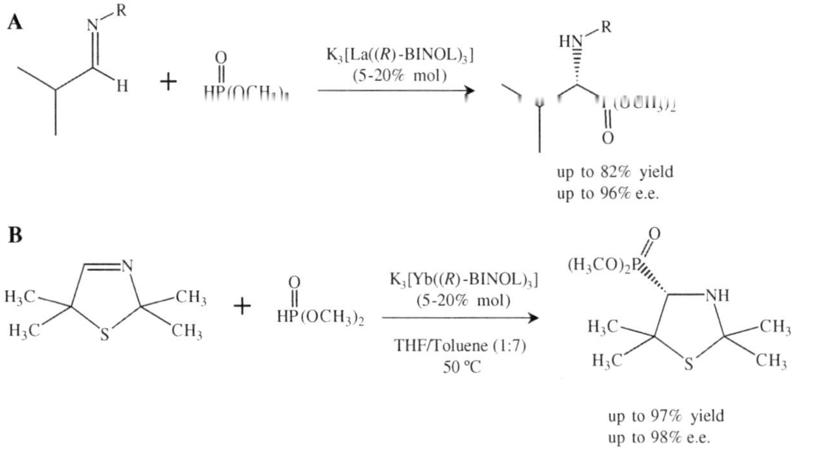

Figure 4.8. Enantioselective hydrophosphonylation of acyclic imines (**A**) and thiazolines (**B**).

The products are chiral α-hydroxy and α-amino phosphonates, which recently met a wide interest in the synthesis of biologically active enzyme inhibitors [16, 17]. In the imine hydrophosphonylation, the potassium systems are the more active catalysts (yields and e.e. over 90-95%), where sodium and lithium complexes still produce high e.e. but much lower yields (generally < 55%). The complexes of the early lanthanides (as $K_3[La((S)\text{-BINOL})_3]$) give efficient hydrophosphonylation on acyclic imines (Figure 4.8A [12], at variance, the hydrophosphonylation of thiazolines (Figure 4.8B) is much better catalysed by smaller lanthanides (Gd, Dy, Yb), and the ytterbium complex $K_3[Yb((S)\text{-BINOL})_3]$ (2) is the most efficient [18, 19].

The mechanism of the thiazoline hydrohosphonylation catalysed by 2 was studied in detail (Figure 4.9) [19, 20].

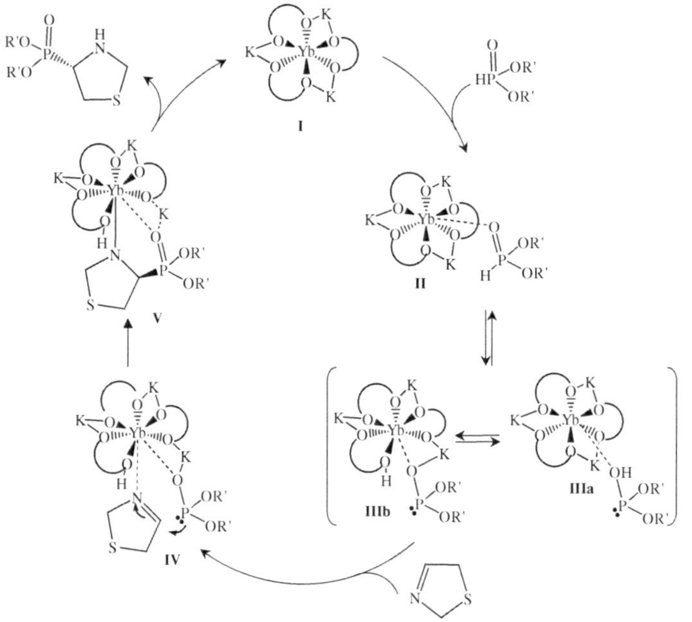

Figure 4.9. Catalytic cycle proposed for the enantioselective hydrophosphony-lation of thiazolines. For the sake of simplicity, the substituents on the thiazolin ring are not reported.

The phosphonate is the first coordinated substrate, because of the oxophilicity of the Ln^{3+} ions, as supported by ^{31}P NMR evidences [19]. To generate the nucleophilic phosphorus for the imine insertion (**IIIb**), the phosphonate/phosphite tautomerism must be invoked (equilibrium between **II** and **IIIa**). Gröger et al. demonstrate that thiazolines must be activated by a Lewis acid to allow the phosphite insertion [19]; for this reason, coordination to the lanthanide in the intermediate **IV** is postulated. In the last step, the proton exchange between the product and the complex regenerates the catalytic species (**I**).

It is supposed that the phosphonate/phosphite tautomerism is the slow passage in the catalysis. This step is promoted by the potassium complexes better than by other alkali metal, probably in view of the higher basicity of the potassium binaphtholate (ROK > RONa > ROLi) [20]. Introducing substituents on the ligand (*i.e.* 6-6′ disubstitued binaphtholate) [21] or changing the alkyl groups of the phosphonate (*i.e.* using a cyclic phosphonate) the catalytic efficiency improves, reaching e.e. up to 98-99% [20]. The rigid conformation of cyclic phosphonates may reduce the activation entropy in the coordination and promote phosphonate enolization through anomeric effects, favouring the formation of the nucleophilic phosphite [20].

4.2. The synthesis of Ln-heterobimetallic complexes

4.2.1. State of art

The first lanthanide heterobimetallic catalyst that appeared in the literature was $Li_3[La((S)$-BINOL$)_3]$ [1]; it was synthesised from La(t-BuO)$_3$ (1 mol eq.), (S)-BINOL-H$_2$ (1.5 mol eq.), LiCl (2 mol eq.) and H$_2$O (10 mol eq.) (Figure 4.10). The complex was prepared in THF and directly used for the catalysis.

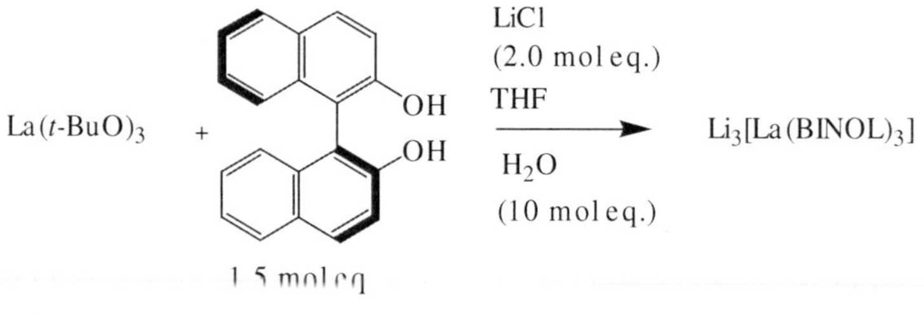

Figure 4.10. First synthesis of heterobimetallic catalyst.

Since the stoichiometry of the catalytic complex became clearer, the procedure was optimised and two different synthetic approaches were devised (Figure 4.11). The first (**A**) [9] starts from the inexpensive LaCl$_3$·7 H$_2$O, dispersing it in a THF solution of lithium binaphtholate (2.7 mol eq.) and in presence of a base (NaOtBu or NaOH 0.3 mol eq.) [22].[3]

---

[3] The base is added to promote the disgregation of the lanthanide chloride. Using dry LaCl$_3$, small amounts of water (about 10 mol equivalents) must be added together with the base. The relative amount of lithium binaphtholate and base reported in literature is variable ([22]).

The second (**B**) starts from the more expensive La(iPrO)$_3$: the complex is obtained stirring the salt in a THF solution of (S)-BINOL-H$_2$ (3.0 mol eq.), in presence of 3.0 mol eq. of a strong base (as BuLi) [7, 9], after 14 h of stirring 1 eq. of H$_2$O was added [7]. By changing the nature of the base (*i.e.* NaOtBu, KN{Si(CH$_3$)$_3$}$_2$), the sodium and the potassium complexes are obtained [11].

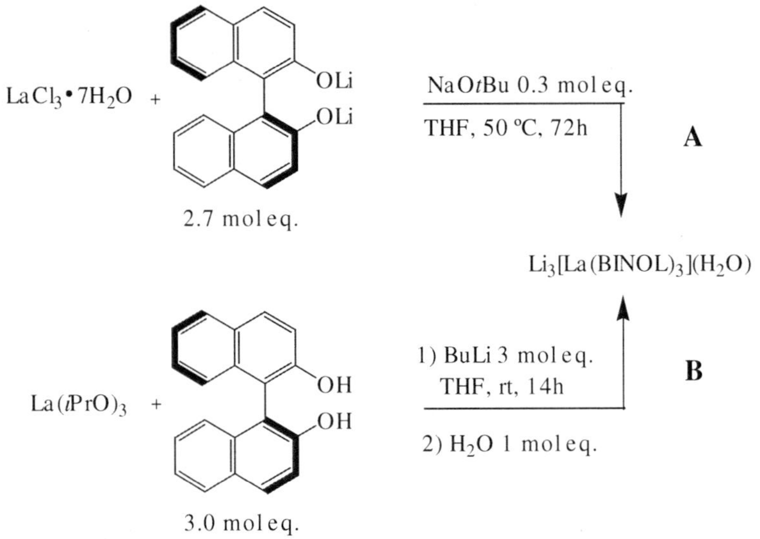

Figure 4.11. Synthetic procedure proposed by Shibasaki for the synthesis of heterobimetallic complexes. **A**) starting from the rare earth halide, **B**) starting from the rare earth isopropoxides.

Independently, Aspinall and co-workers developed a different synthetic route to produce these complexes under anhydrous conditions (Figure 4.12) [23].

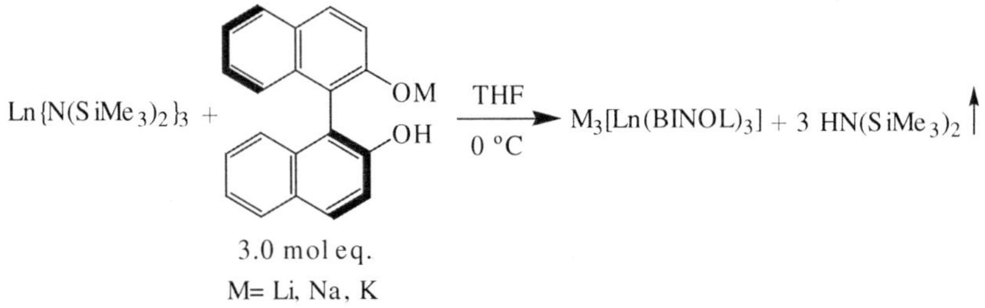

Figure 4.12. Synthetic pathway for the synthesis under anhydrous conditions.

The only by-product of this synthesis is HN(SiMe$_3$)$_2$ that is easily removed under vacuum. This synthesis is completely free from alkali chlorides; this makes easier the crystallization of the complex. In such a way, it was possible to crystallize the lithium derivatives, which are hardly crystallized through the previous procedures [24].

4.2.2. New synthetic route

In this work, we introduced a different synthetic procedure, that starts from the $Yb(OTf)_3$ (Figure 4.13). This procedure was used in the synthesis of the ytterbium sodium and potassium complexes $Na_3[Yb((S)-BINOL)_3]$ (**1**), $K_3[Yb((S)-BINOL)_3]$ (**2**) [25].

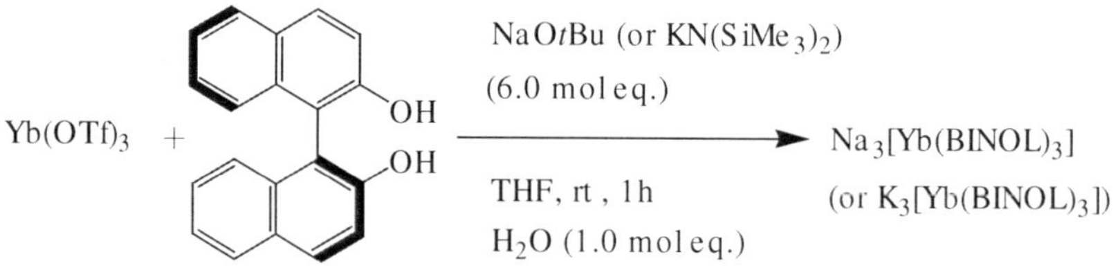

$$Yb(OTf)_3 \ + \quad \text{(BINOL, 3.0 mol eq.)} \quad \xrightarrow[\substack{THF, rt, 1h \\ H_2O \ (1.0 \ mol \ eq.)}]{\substack{NaOtBu \ (or \ KN(SiMe_3)_2) \\ (6.0 \ mol \ eq.)}} \quad Na_3[Yb(BINOL)_3] \\ (or \ K_3[Yb(BINOL)_3])$$

Figure 4.13. New synthetic pathway using $Yb(OTf)_3$.

This synthesis resulted advantageous because the $Yb(OTf)_3$ is completely soluble in THF, not expensive and it is commercially available as anhydrous salt. Moreover, by changing the nature of the base, sodium or potassium complexes are obtained. Finally, the complete absence of alkali halides allowed the easy crystallization of the complexes [11] through a slow addition of *n*-hexane to the complex solution; the crystals obtained in this way have the quality and the dimension for X-ray structure determination [25].

4.3. Crystallographic structures of heterobimetallic complexes

The first crystallographic structures reported for lanthanide heterobimetallic complexes were those of the sodium species $Na_3[Pr((S)-BINOL)_3]$, $Na_3[Nd((S)-BINOL)_3]$ and $Na_3[Eu((S)-BINOL)_3]$ [7]. Later on, the X-ray structure of many other complexes have been reported: for $Ln = La$ ($M = Na$) [11], $Ln = Sm$ ($M = Li$) [5], $Ln = Eu$ ($M = Li$) [23], $Ln = Yb$ ($M = Li$ [23] and Na [23, 25]), and also for yttrium analogues $Ln = Y$ ($M = Li$ [4] and Na [23]). Catalytic tests demonstrate that the solutions obtained dissolving crystals are catalytically active [9, 19], and both NMR [19] and mass spectroscopy (FAB MS, ESI MS [19], and LDI-TOF MS [7]) showed that these complexes are present as single species, which does not dissociate in solution.

In the present thesis, we determined the X-ray structure of **1** · 6THF: it shows a homoleptic coordination of the lanthanide, where three binaphtholate units are symmetrically disposed around the metal, endowing the complex with a main C_3 axis, and reminiscent of the shape of a pinwheel (Figure 4.14). Each alkali metal is bound to two distinct binaphtholic units and to two THF molecules. The octahedral coordination around the lanthanide allows for the Λ/Δ chirality, which combined with the S chirality of the BINOL gives rise to two possible diastereomeric arrangements. In spite of this, in the solid-state[4] and in solution only one species is observed, which has Λ chirality (using (S)-BINOL) [9, 19]. Apparently, the sterical hindrance of the BINOL favours the Λ form that has the 1,1′-intranaphtholic axes oriented almost parallel to the C_3 axis (Figure 4.14 and Figure 4.15).

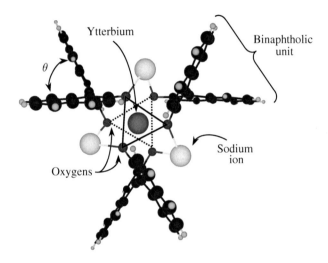

Figure 4.14. Model of the solid state structure of $Na_3[Yb((S)\text{-BINOL})_3]\cdot 6THF$. The THF molecules coordinated to sodium are omitted for sake of simplicity.

The crystal structures of the complexes of the beginning of the series (La, Pr, Nd, Eu) [7, 11] show a water molecule axially coordinated that is absent in the structure of the smaller lanthanides (Yb, Y) [23]. Europium is at the borderline between these two groups: crystal structures of Li_3 $[Eu((S)\text{-BINOL})_3]$ with or without a coordinated water are available [23].

We found that **1** crystallize without coordinated water even in presence of small amounts of water in the mother solution (up to 3 equivalent) [25]. In this complex, Yb has only CN $= 6$ that is much smaller than the usual

---

[4] X-ray diffraction of the crystals powder was performed to rule out the presence of crystals of a second species.

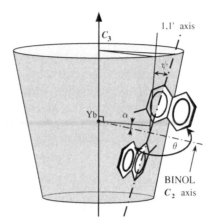

Figure 4.15. Schematic view of the ligand arrangement, the oxygen atoms are omitted for clarity.

CN $= 8$ or 9, probably as a consequence of the large sterical hindrance of the binaphtholate.

In spite of the high symmetry expected for **1**, the crystal structure shows several distortions. The arrangement of the ligands, reproduced in Figure 4.15, shows that the C_2 symmetry axes of the BINOLs are not perpendicular to the principal C_3 axis but form an angle α with it, which can be considered a measure of the *conicity* of the structure. This also reflects in Ln−O bond lengths that are longer for the naphtholate moiety of the upper hemisphere of the complex than for the Ln−O bonds of the lower hemisphere. This bond-length difference is comparable with the one observed in the structures of the Pr, Nd, and Eu complexes, which are distorted by the coordination of a water molecule (Table 4.1). The *conical distortion*, assumed by the ligands, breaks the D_3 symmetry and reduces the molecular symmetry to the C_3 point group.

Table 4.1. Crystallographic Ln−O bond lengths in different complexes measured in the upper hemisphere of the molecule (first row), and in the lower one (second row).

Na₃[Pr((S)-BINOL)₃]	Na₃[Nd((S)-BINOL)₃]	Na₃[Eu((S)-BINOL)₃]	Na₃[Yb((S)-BINOL)₃]
2.386 Å	2.363 Å	2.312 Å	2.230 Å
2.365 Å	2.338 Å	2.286 Å	2.203 Å

The crystal structures reported for the Li₃[Yb((S)-BINOL)₃] show a much less pronounced conical distortion [23]: it is possible that the alkali metals play a role, but there are not enough X-ray structures to extend the correlation and to confirm this hypothesis.

4.4. Solution structural study

4.4.1. The solution structure of $Na_3[Yb((S)\text{-}BINOL)_3]$

In the previous paragraph it was shown that the crystal structure of **1** is conically distorted, even if there is no water axially coordinated. On the contrary, the ^1H NMR spectrum of **1** shows only six proton resonances that correspond to one naphtholic unit, which indicates that in solution the structure has an *effective* D_3 symmetry [23, 25]. The contrast between the crystallographic structure and the NMR data suggests that, upon dissolution, **1** undergoes a structural rearrangement (*static rearrangement*), or alternatively, that it might be simultaneously present as two conically distorted structures with opposite conicity α in fast exchange with respect to the NMR time scale (*dynamic rearrangement*, Figure 4.16).

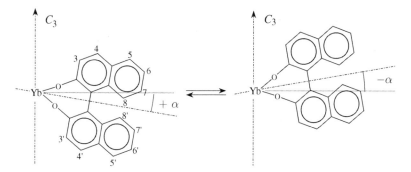

Figure 4.16. Dynamic rearrangement hypothesis for **1**. The exchanging species with conical distortion $+\alpha$ and $-\alpha$ are represented on the left and on the right, respectively.

The hypothesis of *dynamic rearrangement* was investigated analysing the NMR paramagnetic shift (see details in [25]): to have only six proton peaks, the equilibrium of Figure 4.16 must be fast enough to average the resonances of the nuclei of the upper with those of the lower part of the complex (*i.e.* to average the shift of the proton 3 and 3′, 4 and 4′, and so on). As the paramagnetic NMR shift strongly depends on the position of the nuclei (see Chapter 2), the proton shifts in the upper and lower hemisphere are expected to be quite different for a static conformation.

In particular, on the basis of the crystallographic structure we calculate that the protons 3 and 3′ would be separated by more than 15 ppm (equivalents to more that 4500 Hz at 300 MHz of ^1H Larmor frequency)! The dynamic rearrangement should be fast enough to average at least these 4500 Hz into a single resonance having the observed experimental linewidth. So, these data, combined with the experimental ^1H NMR linewidth of **1**, made possible to fix the lower limits for exchange rate in Figure 4.16 to $k > 10^7 \, s^{-1}$. This kinetics rate is fast enough to reasonably exclude the presence of such dynamics [25]. Furthermore, even lowering the temperature down to $-100\,°C$, no decoalescence of the proton resonances is observed, confirming again that a real *static* rearrangement occurs upon dissolution, and an accurate determination of the solution structure is needed.

The solution structure was calculated on the basis of paramagnetic NMR data of **1** (shifts and relaxation rates), and using the computer program PERSEUS (Section 2.4.1). PERSEUS rearranges an initial starting model in order to get the best match between the experimental paramagnetic data[5] and the δ^{pc} e and T_1^{para} calculated from (2.25) and (2.48-50), The crystallographic structure of **1** was used as starting model.[6] In the structural optimisation, PERSEUS rearranges the position of the binaphtholate with respect to Yb (3 variables) and the amplitude of the internaphtholic dihedral angle θ (1 variable) (Figure 4.15); the geometry of each naphtholic moiety remains unchanged with respect to the starting model.[7] The number of experimental constraints (6 ^1H paramagnetic shifts, 6 ^{13}C paramagnetic shift and 6 ^1H T_1) is large enough to fit all the structural variables (4), and the paramagnetic parameters (3 parameters for the magnetic anisotropy tensor, and 1 parameter for the relaxation rates), (further details in [25]).

In Table 4.2, the data fitted on the crystallographic and the rearranged solution structures are compared. The ligand conformations are described by three angles: α, which measures the conical distortion, θ, the internaphtholic dihedral angle, and ψ, which is the dihedral among the C_3 axis, the binaphtholic C_2 axis and the intranaphthalenic 1-1′ axis (Figure 4.15). The $R(\delta^{pc})$ and $R(\rho^{para})$ agreement factors[8] for the shift and relaxation rates, respectively, indicate that the rearranged structure gives a much better agreement with the experimental NMR data. This can also be shown comparing the experimental and calculated NMR spectra (Figure 4.17).

---

[5] The experimental paramagnetic shifts are determined through (2.1) and (2.2) using the shift of the Na$_3$[Lu((S)-BINOL)$_3$] as diamagnetic reference and neglecting the contact contributions.

[6] The PERSEUS calculations were conducted in this case with and without correction for the C−H bond-length to 1.10 Å, with no appreciable differences.

[7] Only the conformational degrees of freedom are changed (see Section 2.4.1).

[8] These agreement factors are defined following the equations (2.53) and (2.54), where $\delta^{pc(calc)}$ and $\rho^{para(calc)}$ are the mean values calculated over all the equivalent nuclei.

Table 4.2. Conformational and magnetic parameters for the crystallographic and solution structures.

Optimised Parameters	Crystallographic Structure	Solution Structure
$\mathcal{D}$	(1900 ± 60) ppm Å^3	(1770 ± 50) ppm Å^3
θ	$(63 \pm 1)°$	$(73 \pm 2)°$
ψ	$(11 \pm 1)°$	$(19 \pm 2)°$
α	$(5 \pm 1)°$	$0°$
$R(\delta^{pc})\%$	13%	5.5%
$R(\rho^{para})\%$	1.8%	3.7%

The solution structure is D_3 symmetric ($\alpha = 0$) with the *intra*-binaphtholic dihedral angle θ about $10°$ larger than in the crystal structure and with some minor variation about the angle ψ (Figure 4.18 and Table 4.2). The increased angle θ accounts for the flexibility of the BINOL [26], and it is not surprising that in solution, with an increased mobility of the ligand, this angle changes to accommodate the structural rearrangement. Even comparing θ in different crystallographic structures, one can observe that θ changes over more than $11°$ along the lanthanide series allowing the ligand to adapt to the variations of the cation size [23].[9]

The complex **1** remains uncoordinated by water in solution. Indeed we observe that upon addition of a small amount of water to a solution of **1** (up to 3 equivalents), the H_2O resonance is neither shifted nor a new peak of bound water is observed, demonstrating that there is no interaction with the paramagnetic center.

4.4.2. Accuracy of the solution structure of Na₃[Yb((S)-BINOL)₃]

The accuracy in the atomic position of the solution geometry is not the same for all the nuclei in the molecule: indeed, pseudocontact shifts and relaxation rates depend on r^{-3} and r^{-6}, respectively ((2.25) and (2.48-50)), and are much more sensitive to variations affecting the nuclei close to metal than that far from it. On the other side, the δ^{para}'s of nuclei closer than three bond lengths from the metal may be affected by contact contributions. In **1**, almost all the nuclei used in the fitting are more than three bonds lengths from Yb (only the C-3 is three bond lengths) and the contact term may be neglected (as discussed in Section 4.4.3).

---

[9] Aspinall *et al.* reported that MM2 calculation done on Li₃[Y((R)-BINOL)₃] produced *intra*-binaphtholic θ angles $12°$ larger than the one observed in the XRD structure, in good agreement with what found in solution.

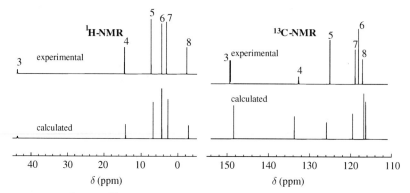

Figure 4.17. ^{1}H and ^{13}C spectra reconstructed on the basis of the calculated and experimental data.

Furthermore, BINOL is a rigid ligand with only one degree of freedom (the rotation around the 1-1' bond), so the error induced by the ligand mobility is reduced.

All these considerations may explain the very good agreement observed between the calculated and experimental spectra of Figure 4.17

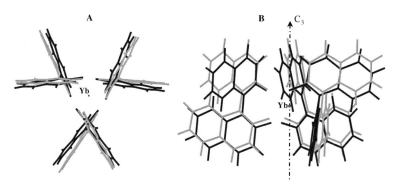

Figure 4.18. Superimposition of crystallographic (grey) and solution structure (blue). For sake of clarity, only the binaphtholate groups (neglecting the oxygen atoms) are reported. Notice the difference in the dihedral θ, and in the tilt angle α in the projections from the top (**A**) and from one side (**B**).

To estimate the error in the structural parameter optimised by PERSEUS, one can observe the plot of the agreement factor $R(\delta^{pc})$ (or $R(\rho^{par})$), when the optimised variables are moved around their minima. Figure 4.19 reports the shift agreement factor $R(\delta^{pc})$ for structures optimised with different angles θ. The minimum is found for $\theta = 73.5°$ but all the angles in the range $73 \pm 2°$ can be accepted as well, as they produce a variation of $R(\delta^{pc})$ within 5% (which is the same order of uncertainty of the NMR data).

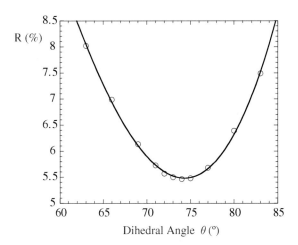

Figure 4.19. Plot of the agreement factor $R(\delta^{pc})$ calculated for solution structures with different *inter*-naphtholic dihedral angle θ. The $R(\delta^{pc})$'s are calculated optimising solution structures with fixed θ angles.

It should be noticed that the angle θ is determined with a larger error with respect to α (Table 4.2): this is essentially due to the cylindrical symmetry of the pseudocontact interaction.[10] Indeed, as the pseudocontact shift (2.25) depends only on the radius r and the azimutal angle,[11] the NMR shift is much more sensitive to a displacement along α, which produces a direct variation in the azimutal angle, than in θ, which affects r and the azimutal angle marginally. This example shows well how the resolution in the solution structure cannot be uniform for all nuclei in the molecule.

4.4.3. Comparing the solution structures of Na$_3$[Yb((S)-BINOL)$_3$], Li$_3$[Yb((S)-BINOL)$_3$] and K$_3$[Yb((S)-BINOL)$_3$]: the role of the alkali metal

The ^{1}H NMR spectra of the complexes K$_3$[Yb((S)-BINOL)$_3$] (2) and Li$_3$[Yb((S)-BINOL)$_3$] (3) show only six resonances as 1. Taking as diamagnetic reference the lutetium complex Na$_3$[Lu((S)-BINOL)$_3$] (4), the paramagnetic shift of 1, 2 and 3 are plotted against the geometrical factor ($GF = (3\cos^2\theta - 1)/r^3$ see (2.25)) of the solution structure of 1 (Figure 4.20).

---

[10] Minor contributions, derived from the BINOL flexibility, can not be completely excluded.

[11] The azimutal angle is the angle θ in Figure 2.3 and (2.25) that should not be confused with the intra-binaphtholic dihedral angle that, in this chapter, is also called θ.

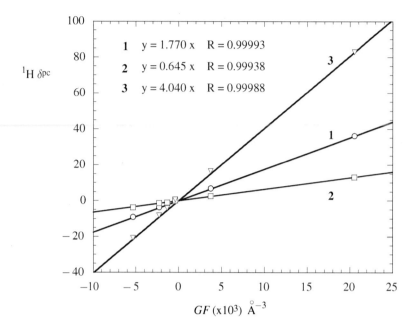

Figure 4.20. Plot of the ^{1}H pseudoconact shift of the complexes **1**, **2**, and **3**, against the geometrical factors (GF) of the solution structure of **1**.

The straight lines obtained in all the cases demonstrate that the solution structure optimized for **1** is compatible also with the data for complexes **2** and **3**. The differences in the NMR spectra depend on the different anisotropy susceptibility factors $\mathcal{D}$ that scales proportional shifts for the three complexes [25].[12] As observed for **1**, the ^{1}H NMR spectrum shows that also **2** is not coordinated to water in solution [25].

The linearity of the paramagnetic shift as a function of the GF's (Figure 4.20) clearly shows that the pseudocontact interaction dominates almost entirely the paramagnetic shift: indeed, contact contribution can not be proportional from a complex to the other, and it must be negligible with respect to δ^{pc}.[13]

---

[12] Strictly speaking, Figure 4.20 demonstrates that the geometries of **1**, **2** and **3** have the same GFs. In general, it is possible that different structures have the same GF's. This is the case of the conformers of the Ln[DOTA] complex (see Chapter 5), but in the present case we found it unrealistic, and **1**, **2** and **3** can be surely assigned to the same structure.

[13] For the sake of precision, plotting the ^{13}C paramagnetic shifts of **2** against the GF of **1**, the carbon 3 does deviate from the line, probably because of contact contribution. This is not surprising as it is only three-bonds away from Yb.

In contrast with the solution, the crystallographic structures **1** and **3** are not equivalent (Section 4.3), and their distortion can be related to the nature of the alkali metals. In the crystal is possible that the ligands are partially constrained and assume different conformations with respect to the solution: so, passing from **3** to **1**, the larger ionic radius of Na^+ produces a stronger conical distortion [23].

Following Bleaney's theory, [27, 28] the variation of the $\mathcal{D}$ factor in **1**, **2**, and **3** is connected to the different crystal-field around the metal. As the structure is the same for all these complexes, we suppose that the crystal field variation is due to the change in the charge density on the donors. Probably, the alkali metals M differently polarizes the M−O bond and induces different charges on the oxygen, which in turn influences the crystal field on Yb.

4.5. NIR CD solution study

4.5.1. Comparison between Na$_3$[Yb((S)-BINOL)$_3$] and K$_3$[Yb((S)-BINOL)$_3$]

As described in Chapter 3, NIR CD spectroscopy reflects the distortion of the coordination polyhedron around ytterbium, and is extremely sensitive to the changes in the arrangement and charge distribution of the donors. Thus, the analysis and the comparison of the NIR CD spectra of different complexes allows one to obtain indications about the geometries around the metal.

The NIR absorption and CD spectra of **1** and **2** show a remarkable correspondence in the sign and intensity of the bands (Figure 4.21), but a different spectral width (larger for **1** than for **2**), which produces the apparent frequency shift between the two spectra. The analogy between the NIR CD spectra confirms again that the two complexes have the same coordination polyhedron, in agreement with what observed through NMR.

Instead, the difference in the spectral width indicates that the two complexes have different crystal-fields around the metal, which induces distinct separation in the F-levels of Yb. Then, three indications are derived from the NIR CD spectra:

i) **1** and **2** have the same solution structure.
ii) **1** and **2** have the same CN and both remain uncoordinated by the solvent.
iii) **1** and **2** have a different charge distribution, probably due to the different polarization of the M−O bond.

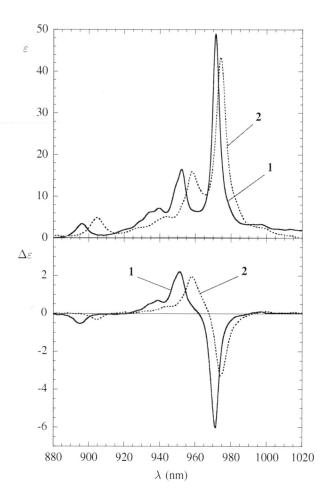

Figure 4.21. Absorption and NIR CD spectra of **1** (solid line) (12.6 mM) and **2** (dotted line) (19.4 mM) in dry THF.

4.5.2. Analysis of the NIR CD spectrum

Extracting structural information from the NIR CD spectrum requires a detailed analysis, in order to correlate intensities, width and frequencies of the bands with the arrangement of the coordination polyhedron, hence with the crystal field parameter. This is a very complicated matter, that can not be completely developed within the present work, we limit ourselves to tentatively assign the observed bands to electronic transitions, which is the first step in the interpretation of the whole spectrum.

This analysis was successfully achieved for the macrocyclic [Yb((R)-DOTMA)] complex, [29] where all the 12 possible transitions were resolved and assigned (Figure 3.4). In many other complexes the overlap among the bands prevents one to resolve all the transitions, hampering the spectral assignment.

Keeping in mind the scheme of electronic levels of Figure 3.3, the transition at shorter wavelength (895 nm for $Na_3[Yb((S)-BINOL)_3]$ and 905 nm for $K_3[Yb((S)-BINOL)_3]$) can be reasonably assigned to the higher energy transition that starts from the lower-lying level of the ground state (1-3' in Figure 3.3). Notably, no other transition are observed up to 935 nm: even assigning this band to the transition starting from the second level of the $^2F_{7/2}$ manifold (2-3'), one must conclude that the first two sublevels of the ground state are separated by at least 290 cm^{-1} (for **2**). Assuming a Boltzmann statistic, the population of the second level at room temperature is lower than 25% of the first one, and consequently, the "hot bands" originating from the levels 2, 3, 4, are expected to be weak. So, it is not surprising that many bands are hardly discerned and often overlapped with the stronger ground state transitions.

Lowering the temperature, two distinct effects are expected in the NIR spectra:

i) the increase of the Rotational Strength for the transitions starting from the sublevel 1 due to its increased population;
ii) the reduction of the bands linewidth, with the consequent enhancement of the bands resolution.

In Figure 4.22, the spectra of the complex $K_3[Yb((S)-BINOL)_3]$ recorded at 298 K and 193 K are compared: almost all the bands are evidently sharper and resolved at low temperature, and other 6 weak bands and shoulders are identified in addition to stronger bands at 905, 958, 974 nm. To obtain the Rotational Strengths for each transition (3.4), the spectra at 298 K and 193 K were deconvoluted as a sum of 9 single peaks centered at 905, 930, 944, 953, 958, 974, 983, 998, 1015 nm.[14] The bands at 930, 944, 953, 983, and 1015 nm are largely overlapped and have integrals much smaller than the bands at 905, 958, and 974 nm, so their quantitative measure is not considered because is too error prone. Only the bands at 905, 958, 974, and 998 nm are intense enough or resolved to be accurately integrated. Considering Lorentzian lineshapes, equation

---

[14] For each transition, a Lorentzian lineshape in the wavenumber domain was assumed. The peak intensities were obtained by fitting the experimental spectra with the function $y = \sum_{i=1}^{9} \Delta\varepsilon_i^{max} / (1 + ((x - \tilde{\nu}_i)/\Delta\tilde{\nu}_i)^2)$, where $\Delta\varepsilon_i^{max}$ is the maximum peak intensity, $\tilde{\nu}_i$ in the peak wavenumber and $2\Delta\tilde{\nu}_i$ is the half-height peak linewidth.

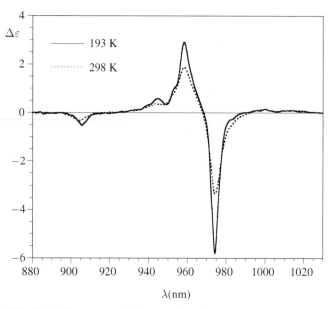

Figure 4.22. NIR CD spectra of the complex **2** recorded at 298 K (dotted line) and 193 K (solid line).

(3.4) simplifies into

$$
\begin{aligned}
R_{ij} &= \frac{3hc10^3 \ln 10}{32\pi^3 N_A} \int \frac{\xi_i \Delta\varepsilon_{ij}}{\tilde{v}} d\tilde{v} \cong \frac{3hc10^3 \ln 10}{32\pi^3 N_A} \frac{\xi_i}{\tilde{v}_{ij}} \int \Delta\varepsilon_{ij} d\tilde{v} \\
&= \frac{3hc10^3 \ln 10}{32\pi^3 N_A} \frac{\pi \xi_i \Delta\varepsilon_{ij}^{\max} \Delta\tilde{v}_{ij}}{\tilde{v}_{ij}} .
\end{aligned}
\tag{4.1}
$$

If $\Delta\varepsilon$ is expressed in M^{-1} cm^{-1}, and the wave number in cm^{-1}, in the cgs units the Rotational Strength becomes

$$
R_{ij} = 2.297 \times 10^{-39} \frac{\pi \xi_i \Delta\varepsilon_{ij}^{\max} \Delta\tilde{v}_{ij}}{\tilde{v}_{ij}} .
\tag{4.2}
$$

Using the practical Debye-Bohr magneton units (1 Debye (D) $= 10^{-18}$ esu, and 1 Bohr magneton (μ_B) $= 9.27 \ 10^{-21}$ erg gauss^{-1}), (4.2) becomes

$$
R_{ij} = 0.248 \frac{\pi \xi_i \Delta\varepsilon_{ij}^{\max} \Delta\tilde{v}_{ij}}{\tilde{v}_{ij}} .
\tag{4.3}
$$

Comparing the rotational strength of the bands at 298 and 193 K (Table 4.3), it is apparent that these values increase at low temperature for the bands at 905, 958, and 974 nm, and reduces for that one at 998 nm.

Table 4.3. Intensity and Rotational Strength $(R(T))$ calculated for the four main transitions in Figure 4.22 at 298 K and 193 K. The Corrected Rotational Strengths (R') are calculated from (4.5) and expressed in Debye-Bohr magneton units $(D\,\mu_B)$.

λ	Transition	$T = 298$ K			$T = 193$ K		
(nm)		$\Delta\varepsilon^{\max}(\pm 0.01)$ $M^{-1}cm^{-1}$	$R(T)(\times 10^3)$ $D\mu_B$	R' $(\times 10^3)$	$\Delta\varepsilon^{\max}(\pm 0.01)$ $M^{-1}cm^{-1}$	$R(T)(\times 10^3)$ $D\mu_B$	R' $(\times 10^3)$
905	1 3′	0.34	2.24	3.21	0.52	3.00	3.50
958	1 2′	−2.09	−16.7	−23.9	−2.97	−18.6	−21.7
974	1 1′	3.52	17.9	25.6	6.00	20.5	23.9
998	2 1′ (3 2′)	−0.19	−2.04	−11.9	−0.15	−0.779	−7.9

This indicates that the three more intense bands are those starting from the level 1 in the $^2F_{7/2}$ set, which becomes more populated on lowering the temperature. Reasonably, the bands at 939, 944, 953 nm can be assigned to the transitions 2-3′, 3-3′, 4-3′, respectively, as reported in Figure 4.23 and Figure 4.24. The assignment of the spectra of **1** is performed through the simple comparison with the assignment of **2** at 298 K.

In these tentative assignments several transitions fall at wavelength higher than 1000 nm (< 10000 cm^{-1}), where only very weak bands, hardly distinguished from the baseline, are observed (Figure 4.23). These transitions are those starting from the levels 3 and 4 of the $^2F_{7/2}$ set, which are much less populated than the levels 1 and 2. In Table 4.4 the Boltzmann populations $b(T)$ (4.4) at 298 K and 193 K expected from the level diagram in Figure 4.24 are reported. It is deduced that the level 4 has $\sim 5\%$ of the population of the level 1 at 298 K (1.5% at 193 K), which reasonably justifies the observed weak Cotton effects (Figure 4.23).

$$b_i(T) = \frac{e^{-E_i/kT}}{\sum_i e^{-E_i/kT}} \tag{4.4}$$

$$R'_{ij} = R_{ij}/b_i(T). \tag{4.5}$$

Table 4.4. Boltzamann polulation $(b(T))$ for the levels of the $^2F_{7/2}$ set at 298 K and 193 K for the complex **2**.

SUBLEVEL	b (298 K)	b (193 K)
1	0.698	0.857
2	0.172	0.099
3	0.079	0.030
4	0.051	0.015

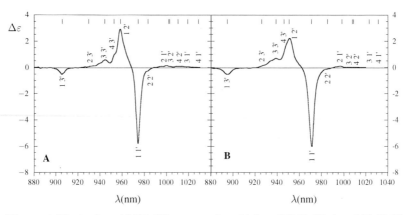

Figure 4.23. Assigned NIR CD spectra for: **A) 2** at 193 K, **B) 1** at 298 K. The short ticks indicate the position of the predicted transitions.

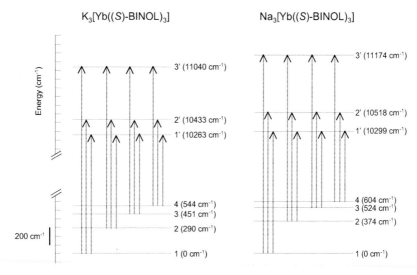

Figure 4.24. Electronic level diagrams for the F states of Yb(III) in the complexes **2** (left side) and **1** (right side). Notice the stronger crystal field split of the levels in **1** with respect **2**.

Correcting the Rotational Strengths for the Boltzamann factor $b(T)$ (4.4) (R'_{ij} in Table 4.3), we observe a good agreement among the values measured at 298 K and those at 193 K. The band at 998 nm, assigned to two the transition 2-1′, has a partial overlap with the transition 3-2′. Consequently, the temperature dependence of this band can not be easily predicted on the basis of (4.4), and the corresponding R'_{ij} shows a significant deviation.

The level diagrams of Figure 4.24 evidence that the crystal-field interaction produces a larger splits of the $^2F_{7/2}$ and $^2F_{5/2}$ state sets in **1** with respect to **2**; the center of gravity of all the transitions is around 980 nm for both the complexes, in agreement with the theory (Section 3.1.2). Consequently, the spectral widths span from 890 to 1035 nm for **1** against 900-1025 nm for **2**. As introduced in the previous paragraph, this difference in the crystal-field splitting may arise from a different charge distribution on the donors, in turn related to the different polarization of the Na−O and K−O bonds.

4.6. Ligand exchange

In order to pose stronger bases for the rationalization of the catalytic behaviour of the $M_3[Ln(BINOL)_3]$ family, the solution study was enriched with the investigation of the ligand lability. In particular, we studied the ligand-exchange in $K_3[Yb((R)\text{-BINOL})_3]$ ((R)-**2**) upon addition of an excess of the free ligand (R)-BINOL-H_2.[15]

The ^1H NMR spectrum of an equimolar solution of (R)-**2** and (R)-BINOL-H_2 is the exact the superposition of the resonances of the two compounds without appreciable linewidth increase, but the 2D EXSY experiment clearly evidences a slow exchange between free and bound BINOL (Figure 4.25).

As described in the equilibrium [4.1], the ligand exchange corresponds also to a proton exchange between free binaphthol (BINOL-H_2 = LH_2) and binaphtholate (BINOL = L).

$$K_3[Yb((R)\text{-L})_3] + (R)\text{-LH}_2 \underset{k_{-1}}{\overset{k_1}{\rightleftharpoons}} K_3[Yb((R)\text{-L})_3] + (R)\text{-LH}_2 \quad [4.1]$$

Through the analysis of the build up in NMR saturation transfer experiment, the global kinetic constant for the equilibrium [4.1] was measured as $k_1 = (370 \pm 30)$ mol^{-1}s^{-1} (see [30] for details). Moreover, the NMR spectra do not show any evidence of intermediate species.

Two hypotheses about the mechanism of exchange can be put forward:

i) the ligand dissociates as alkaline binaphtholate, which rapidly gives rise to the proton exchange in solution [4.2a,b];
ii) the proton exchange and the ligand detachment occur in the coordination sphere of ytterbium in a concerted or pseudo-concerted mechanism.

---

[15] This situation is analogue to that one observed when a large excess of water is added to a solution of (R)-**2** and the partial decomposition of the complex produces free BINOL-H_2.

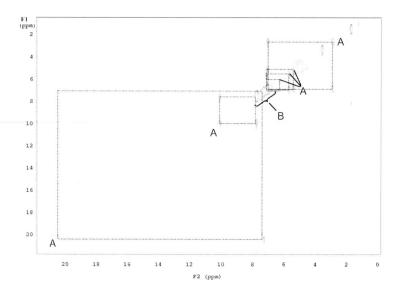

Figure 4.25. EXSY spectrum of (R)-**2** in the presence of about 1 eq. of (R)-BINOL-H_2 in d_8-THF. The resonances of the complex (R)-**2** are labelled A, while the (R)-BINOL-H_2 are B.

$$K_3[Yb((R)\text{-L})_3] \rightleftharpoons K_2[Yb((R)\text{-L})_2]^+ + [(R)\text{-L-K}]^- \quad [4.2a]$$

$$[(R)\text{-L-K}]^- + (R)\text{-LH}_2 \rightleftharpoons (R)\text{-LH}_2 + [(R)\text{-L-K}]^- \quad [4.2b]$$

As (R)-**2** does not appreciably dissociate in solution, [7, 19] is more plausible that the exchange proceed through a concerted mechanism, but sure proofs of this are, up to now, not available. The concerted mechanism might also go through a partial ligand detachment that favours the coordination of the new ligand and the proton exchange.

When the same experiment was done mixing (R)-**2** with the ligand of opposite configuration (S)-BINOL-H_2, the ligand exchange produced a mixture of stereoisomers. In Figure 4.26 the NIR CD spectra of (R)-**2** before, and immediately after the addition of the 1.2 equivalents of (S)-BINOL-H_2 are reported.

It is evident that the ligand exchange also occurs with this ligand and, that after the first substitution, the process proceeds until all the binaphtholates are replaced, giving the homochiral (S,S,S)-**2** species (or simply **2**). Apart from sign, the spectra in Figure 4.26 are very similar to each other, with no indication for intervening heterochiral intermediates (S,R,R)-**2** and (S,S,R)-**2**. On the contrary, when the same experiment

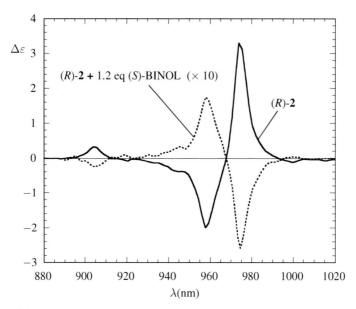

Figure 4.26. NIR CD spectra of the complex (R)-**2** (18.0 mM) before, and after the addition of 1.2 equivalents of (S)-BINOL-H_2.

was followed through ^{1}H NMR, we observed a small amount of the heterochiral species, not appreciated in the CD spectra (Figure 4.27). The 2D EXSY confirmed that the six new peaks of the heterochiral species are in slow exchange with both the free BINOL-H_2 and the homochiral forms (*i.e.* (R)-**2** and **2**), as summarized in [4.3a,b,c]. In addition to these equilibria, the homochiral exchange [4.1] is still active between the free ligand and the complex with the same chirality.[16]

$$K_3[Yb((R)\text{-}L)_3] + (S)\text{-}LH_2 \underset{k_{-a}}{\overset{k_a}{\rightleftharpoons}} K_3[Yb((R)\text{-}L)_2(S)\text{-}L] + (R)\text{-}LH_2 \qquad [4.3a]$$

$$\textbf{(I)} \qquad\qquad\qquad\qquad\qquad \textbf{(II)}$$

$$K_3[Yb((R)\text{-}L)_2(S)\text{-}L] + (S)\text{-}LH_2 \underset{k_{-b}}{\overset{k_b}{\rightleftharpoons}} K_3[Yb((S)\text{-}L)_2(R)\text{-}L] + (R)\text{-}LH_2 \qquad [4.3b]$$

$$\textbf{(II)} \qquad\qquad\qquad\qquad\qquad\qquad \textbf{(III)}$$

$$K_3[Yb((S)\text{-}L)_2(R)\text{-}L] + (S)\text{-}LH_2 \underset{k_{-c}}{\overset{k_c}{\rightleftharpoons}} K_3[Yb((S)\text{-}L)_3] + (R)\text{-}LH_2 \qquad [4.3c]$$

$$\textbf{(III)} \qquad\qquad\qquad\qquad\qquad \textbf{(IV)}$$

---

[16] The simultaneous action of so many equilibria gives a reason of the increased linewidth observed in the resonances of **2**.

The heterochiral complexes **II** and **III** form an enantiomeric pair, as well as the homochiral pair **I** and **IV**; thus, their NMR spectra are not distinguishable and only two sets of signals are detectable: those of the pair (**I**, **IV**) and those of the pair (**II**, **III**). The species **II** and **III** show only six broad signals, which are apparently in contrast with the large number of resonances expected for complexes lacking C_3 symmetry. Indeed, ^{1}H NMR spectra of the heterochiral $Li_3[Yb((S)-BINOL)_2((R)-BINOL)]$ and $Na_3[Yb((S)-BINOL)_2((R)-BINOL)]$ (analogues to **III**) show 18 resonances, in agreement with a C_2 symmetry, which is also the case for the crystallographic structure of $Li_3[Y((S)-BINOL)_2((R)-BINOL)]$ [23].

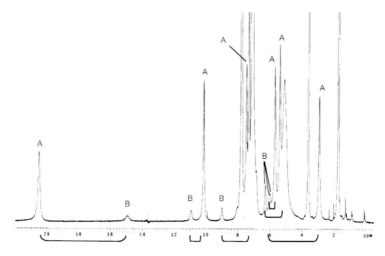

Figure 4.27. ^{1}H NMR spectrum of (R)-2 (11.0 mM) in the presence of 1.0 equivalents of (S)-BINOL-H_2 in d_8-THF (T = 298 K). The square brackets indicate the couples of exchanging peaks between the homochiral species (A) and the heterochiral species (B), as found through EXSY.

Referring to the proposed equilibria [4.3], we should conclude that [4.3b] must be intermediate on the NMR time scale. Indeed, the rapid exchange between the enantiomers **II** and **III** averages the shift of the nuclei of the different BINOL moieties: in this regime only six peaks are expected, which corresponds to the nuclei of a single naphtholic unit.[17]

Unfortunately, as [4.3b] averages the resonances of the heterochiral species, it is not possible to obtain structural information from the analysis of the NMR data. A model for the structure of the complexes **II**

---

[17] The replacement of one (S)-BINOL unit in **II** with the (R)-BINOL (giving **III**) exchanges the shifts of the R and S ligands, because **II** and **III** are enantiomers. If this exchange is fast enough, all the resonances mediate to the six, non-exchangeable, resonances of one naphtholic moiety.

and **III** can be derived from the crystallographic structure of Li$_3$[Y((R)-BINOL)$_2$((S)-BINOL)] (Figure 4.28): the homochiral fragment (*i.e.* the fragment composed of two (R)-BINOL) conserves almost the same arrangement found in the analogous homochiral complexes (as in Figure 4.14). The third ligand ((S)-BINOL), is rotated around its C_2 axis, lying perpendicularly to the other ligands.[18]

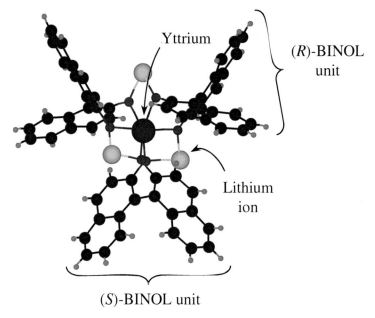

Figure 4.28.   Model of the crystallographic structure of Li$_3$[Y((R)-BINOL)$_2$(S)-BINOL]·6THF. The THF molecules coordinating the sodium ions are omitted for the sake of simplicity.

From the analysis of the NMR peak integrals of **II** and **III** and the intensities of the NIR CD spectrum we estimated the equilibrium constant K_a for [4.3a]: $K_a = (0.13 \pm 0.01)$.[19] It should be noticed that when

---

[18] The combination of the chirality of the ligand, and the chiral arrangement around the metal allows the formation of four stereoisomers (two pairs of enantiomers). On the contrary, both in the crystal structure of Li$_3$[Y((S)-BINOL)$_2$((R)-BINOL)] and the ^{1}H NMR spectra of Li$_3$[Yb((S)-BINOL)$_2$((R)-BINOL)] and Na$_3$[Yb((S)-BINOL)$_2$((R)-BINOL)] only one pair of enantiomers was found [23]. As previously discussed in Section 4.3 for the homochiral complexes, the ligand sterical hindrance is responsible for a strong stereoselection, and the second pair of diastereomers is not formed.

[19] As the chemical potential of enantiomers is the same (in these conditions), we have $K_b = 1$ and $K_a \cdot K_b \cdot K_c = 1$, where K_a, K_b, and K_c are the equilibrium constants for [4.3a], [4.3b], [4.3c], respectively. So, $K_a \cdot K_b \cdot K_c$ is the constant for the whole process of inversion, and consequently $K_a = 1/K_c$.

racemic BINOL is used in the in the synthesis of $Li_3[Yb(BINOL)_3]$ and $Li_3[Y(BINOL)_3]$ the heterochiral species are the species largely prevalent (corresponding to $K_a \gg 1$), in the same conditions the synthesis of $Na_3[Yb(BINOL)_3]$ showed a relative ratio of homochiral/heterochiral of only 3:1 (corresponding to $K_a \simeq 0.33$) [23]. Comparing these values it appears that in racemic mixtures, the relative stability of the heterochiral forms with respect to the homochiral one progressively decreases going from the Li to the K complexes. Aspinall and co-workers suggested that the stabilizing "edge to face" interactions between C−H of the (R)-BINOL and the π system of the adjacent (S)-BINOL's favour the heterochiral arrangement with respect to the homochiral one [23]. It is plausible that increasing the radius of the alkali metal (in combination with the smaller size of the lanthanide ion (*i.e* Y, Yb)), this interaction is hampered and the homochiral arrangement becomes more stable.

4.7. New hypothesis in the catalytic mechanism

In Sections 4.4, 4.5, and 4.6 two important results are introduced:

i) (R)-**2** has a very little tendency to expand its coordination sphere by coordinating other molecules (as water);
ii) the binaphtholate ligands are labile and may rapidly exchange with both (R)-BINOL-H_2 or (S)-BINOL-H_2.

These results would suggest that the catalytic mechanism might proceed through a dissociative initial step instead of the associative step reported in Section 4.1.3 (**II**, Figure 4.9). The associative step was hypothesized once the crystallographic structures of the heterobimetallic complexes were resolved, [7, 9] and the structures (Ln = Pr, Nd, Eu) effectively showed a coordinated water molecule (CN = 7). More recently, the associative mechanism was demonstrated by Shibasaki and coworkers in the enantioselective aza-Michael reaction catalysed by $Li_3[Y((S)-BINOL)_3]$, [31] but this result, as Shibasaki *et al.* said, [31, 32] is not in contrast with a dissociative hypothesis for the Yb-catalysis. Indeed, the Y(III) ion has a ionic radis larger then Yb(III), and it is also demonstrated through NMR spectroscopy that $Li_3[Y((S)-BINOL)_3]$ partially coordinates water in solution, [31, 32] in contrast with the complexes **1** and **2**. Furthermore, the enantioselective aza-Michael reaction has a quite different mechanism from that of thiazoline hydrophosphonylation, and it was observed that the activity of the aza-Michael reaction strongly decreases using Ln-complexes with smaller ionic radius (Tm, Yb, Lu) and $Li_3[Yb((S)-BINOL)_3]$ is virtually inactive [32].

It results that the debate about an associative or dissociative mechanism for the catalysis of **1** and **2** is still open, as Yb has CN = 6, and no evidences of an associative step are reported. For **1** and **2**, the coordination of the phosphonate [19] seems inappropriate, and alternatively we propose the partial (or total) dissociation of the binapholate that creates a coordinative vacancy needed to bind the substrates (Figure 4.29).

Following this hypothesis, the phosphonate binding occurs jointly with the proton exchange with the binaphtholate (analogously to [4.1]), which promotes the formation of the nucleophilic phosphite; successively, it inserts on the imine double bond. The nature of the alkali metal and of the phosphonate [20] might be determinant in this step, influencing the basicity of the binaphtholate and the rate of exchange.

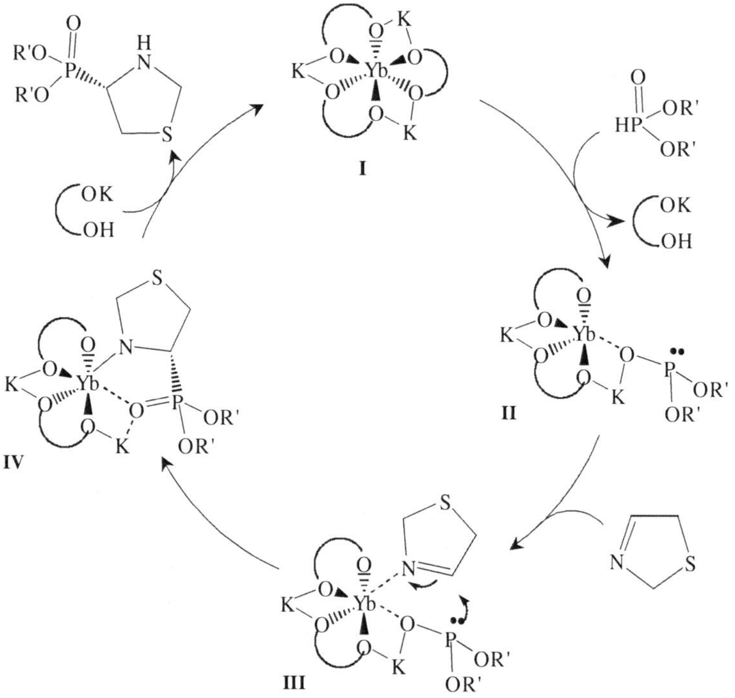

Figure 4.29. New hypothesis of catalytic cycle for enantioselective hydrophosphonylation of the thiazolines. For sake of simplicity, the thiazolin substituents are not reported.

The coordinative vacancy, derived from the ligand detachment, also facilitates the successive thiazoline coordination (**III**). In the last step, the intermediate **IV** may release the product, binding again either the binaphtolate (**IV** → **I**), or a new substrate molecule (following proton exchange)

(**IV** → **II**). In this mechanism only two of the three chiral ligands have an effective influence in inducing the reaction stereoselectivity; this also may support the effectiveness of lanthanide catalysts composed only by two linked BINOL's [33].

At the moment, the dissociative mechanism is only a hypothesis, the fact that Yb has CN = 6 in **1** and **2**, and that BINOL coordination is labile is not sufficient to fully demonstrate a dissociative mechanism; but, if confirmed by further experimental evidences, it should be possible to design, at least for this reaction, new catalysts saving the amount of chiral ligands.

4.8. Conclusions

The work reported in this chapter demonstrates how NMR and NIR CD spectroscopies are valuable techniques in the study of asymmetric ytterbium catalysts. These techniques made it possible to obtain a refined structure of the complex in solution, allowing us to disclose differences with respect to the crystallographic structure. Furthermore, it was possible to discover dynamical processes such as the ligand exchange, and to exclude the coordination of water; all these aspects provide a detailed description of the system in solution, which in turn is an important basis for the investigation of the catalytic mechanism. In detail, the solution structures of the complexes $Na_3[Yb((S)-BINOL)_3]$ and $K_3[Yb((S)-BINOL)_3]$, were determined on the bases of paramagnetic NMR constraints. From these studies it results that the complexes have very similar geometries in solution and, in particular for $Na_3[Yb((S)-BINOL)_3]$, the solution structure differs from the one determined from X-ray diffraction. Moreover, a detailed analysis of the NMR data allows us to reasonably exclude that the solution structure arises from a fluxional dynamics; on the contrary, a true structural rearrangement takes place upon dissolution.

The combined use of NMR and NIR CD spectroscopies made it possible to study the ligand lability process: it was demonstrated that $K_3[Yb((R)-BINOL)_3]$ may rapidly exchange ligand with both (R)-BINOL-H_2 or (S)-BINOL-H_2 giving, in the latter case, the enantiomeric complex $K_3[Yb((S)-BINOL)_3]$ and small amounts of the heterochiral complexes $K_3[Yb((S)-BINOL)_2((R)-BINOL)]$ and $K_3[Yb((R)-BINOL)_2((S)-BINOL)]$.

The possibility of a rapid ligand exchange, together with the observation that $Na_3[Yb((S)-BINOL)_3]$ and $K_3[Yb((S)-BINOL)_3]$ do not bind water in solution and may hardly expand their coordination number, point to revise the proposed catalytic mechanism, introducing a dissociative step preceding or concerned with the substrate binding, at least for ytterbium systems. But more than this, the NMR and NIR CD spectroscopic analyses are demonstrated effective in the study of such molecules permitting to disclose new chemical behaviours. From our first publication introducing the binaphtholic ligand exchange, [25, 30] such dynamics was found involved in several mechanisms of the Ln/BINOL catalysis, [31, 32, 34, 35] and shed new light in understanding the chemistry of the lanthanide heterobimetallic complexes.

Finally, the analysis of the NIR CD spectra of $Na_3[Yb((S)\text{-BINOL})_3]$ and $K_3[Yb((S)\text{-BINOL})_3]$ was performed in detail. The comparison of the spectra at 298 K and 193 K allowed us to assign the more intense bands to their corresponding electronic transitions. These results open the way to develop the correlation between NIR CD spectra and the structure of Yb complexes.

References

[1] SASAI, H.; SUZUKI, T.; ARAI, S.; ARAI, T.; SHIBASAKI, M., *J. Am. Chem. Soc.* **1992**, *114*, 4418–4420.

[2] (a) KOBAYASHI, S., *Lanthanides: Chemistry and Use in Organic Synthesis, Topics in Organometallic Chemistry,* **1999**, Springer-Verlag: Berlin. (b) SHIBASAKI, M.; YOSHIKAWA, N., *Chem. Rev.* **2002**, *102*, 2187–2209.

[3] ASPINALL, H. C., *Chem. Rev.* **2002**, *102*, 1807–1850.

[4] ASPINALL, H. C.; DWYER, J. L. M.; GREEVES, N.; STEINER, A., *Organometallics* **1999**, *18*, 1366–1368.

[5] TAKAOKA, E.; YOSHIKAWA, N.; YAMADA, Y. M. A.; SASAI, H.; SHIBASAKI M., *Heterocycles* **1997**, *46*, 157–163.

[6] EVANS, V.; SOLLBERGER, M. S.; ZILLER, J. W., *J. Am. Chem. Soc.* **1993**, *115*, 4120, and references therein.

[7] SASAI, H.; SUZUKI, T.; ITOH N.; TANAKA, K.; DATE, T.; OKAMURA, K.; SHIBASAKI, M., *J. Am. Chem. Soc.* **1993**, *115*, 10372–10373.

[8] SASAI, H.; TOKUNAGA, T.; WATANABE, S.; SUZUKI, T.; ITOH, N.; SHIBASAKI, M., *J. Org. Chem.* **1995**, *60*, 7388.

[9] SHIBASAKI, M.; SASAI, H.; ARAI, T., *Angew. Chem. Int. Ed. Engl.* **1997**, *36*, 1236–1256.

[10] SASAI, H.; ARAI, T.; SHIBASAKI, M., *J. Am. Chem. Soc.* **1994**, *116*, 1571.

[11] SASAI, H.; ARAI, T.; SATOW, Y.; HOUK, K. N.; SHIBASAKI, M., *J. Am. Chem. Soc.* **1995**, *117*, 6194–6198.

[12] SASAI, H.; ARAI, S.; TAHARA, Y., SHIBASAKI, M., *J. Org. Chem.* **1995**, *60*, 6656–6657.

[13] YOKOMATSU, T.; YAMAGISHI, T.; SHIBUYA, S., *Tetrahedron Asimmetry* **1993**, *4*, 1783.

[14] RATH, N. P.; SPILLING, C. D., *Tetrahedron Lett.* **1994**, *35*, 227.

[15] SASAI, H.; BOUGAUCHI, M.; ARAI, T.; SHIBASAKI, M., *Tetrahedron Lett.* **1997**, *38*, 2717.

[16] KAFARSKI, P.; LEJCZAK, B., *Phosphorus Sulfur Silicon Relat. Elem.* **1991**, *63*, 193–215.

[17] BODUSZEK, B.; OLEKSYSZYN, J.; KAM, C.M.; SELZER, J.; SMITH R. E.; POWER, J. C., *J. Med. Chem.* **1994,** *37,* 3969.

[18] GRÖGER, H.; SAIDA, Y.; ARAI, S.; MARTENS, J.; SASAI, H.; SHIBASAKI, M., *Tetrahedron Lett.* **1996,** *37,* 9291–9292.

[19] GRÖGER, H.; SAIDA, Y.; SASAI, H.; YAMAGUCHI, K.; MARTENS, J.; SHIBASAKI, M., *J. Am. Chem. Soc.* **1998,** *120,* 3089–3103.

[20] SCHLEMMINGER, I.; SAIDA, Y.; GRÖGER, H.; MAISON, W.; DUROT, N.; SASAI, H.; SHIBASAKI, M.; MARTENS, J., *J. Org. Chem.* **2000,** *65,* 4818–4825.

[21] HOU, Z.; ZHANG, Y.; YOSHIMURA, T.; WAKATSUKI, Y., *Organometallics* **1997,** *16,* 2963.

[22] SASAI, H.; SUZUKI, T.; ITOH N.; SHIBASAKI, M., *Tetrahedron Lett.* **1993,** *34,* 851–854.

[23] ASPINALL, H. C.; BICKLEY, J. F.; DWYER, J. L. M.; GREEVES, N.; KELLY, R. V.; STEINER, A., *Organometallics* **2000,** *19,* 5416–5423.

[24] SHIBASAKI *et al.*, published the crystal structure of Li₃[Sm(BINOL)₃] synthesised with the previously reported procedure, but the structure was not well refined (Ref. [5]).

[25] DI BARI, L.; LELLI, M.; PINTACUDA, G.; PESCITELLI, R.; MARCHETTI, F.; SALVADORI, P., *J. Am. Chem. Soc.* **2003,** *125,* 5549–5558.

[26] DI BARI, L.; PESCITELLI, G.; SALVADORI, P., *J. Am. Chem. Soc.* **1999,** *121,* 7998–8004.

[27] BLEANEY, B., *J. Magn. Reson.* **1972,** *8,* 91–100.

[28] (a) MIRONOV, V. S.; GALAYAMETDINOV, Y. G.; CEULEMANS, A.; GOERLLER-WALRAND, C.; BINNEMANS, K., *J. Chem. Phys.* **2002,** *116,* 4673–4685. (b) MIRONOV, V. S.; GALAYAMETDINOV, Y. G.; CEULEMANS, A.; GOERLLER-WALRAND, C.; BINNEMANS, K., *Chem. Phys. Lett.* **2001,** *345,* 132–140.

[29] DI BARI, L.; PINTACUDA, G.; SALVADORI, P., *J. Am. Chem. Soc.* **2000,** *122,* 5557–5562.

[30] DI BARI, L.; LELLI, M.; SALVADORI, P., *Chem. Eur. J.* **2004,** *10,* 4594–4598.

[31] YAMAGIWA, N.; MATSUNAGA, S.; SHIBASAKI, M., *Angew. Chem. Int. Ed.* **2004,** *43,* 4493–4497.

[32] YAMAGIWA, N.; QIN, H.; MATSUNAGA, S.; SHIBASAKI, M., *J. Am. Chem. Soc.* **2005,** *127,* 13419–13427.

[33] (a) KIM, Y. S.; MATSUNAGA, S.; DAS, J.; SEKINE, A.; OHSHIMA, T.; SHIBASAKI, M., *J. Am. Chem. Soc.* **2000,** *122,* 6506–6507. (b) Shibasaki, M.; Matsunaga, S. *Chem. Soc. Rev.* **2006,** *35,* 269–279.

[34] TOSAKI, S.; TSUJI, R.; OHSHIMA, T.; SHIBASAKI, M., *J. Am. Chem. Soc.* **2005,** *127,* 2147–2155.

[35] HORIUCHI, Y.; GNANADESIKAN, V.; OHSHIMA, T.; MASU, H.; KATAGIRI, K.; SEI, Y.; YAMAGUCHI, K.; SHIBASAKI, M., *Chem. Eur. J.* **2004,** *10,* 4594–4598.

Chapter 5
Structure and dynamics of Yb-DOTA-like complexes

The DOTA-like systems are polydentate ligands derived from the cyclen macrocycle (1,4,7,10-tetraazacyclododecane, Figure 5.1). DOTA (1,4,7,10-tetraazacyclododecane-1,4,7,10-tetraacetic acid) is probably the most famous member of this family, and contains four carboxilate groups in the side arms. This branched structure provides eight donors (four oxygens and four nitrogens) and is particularly efficient for lanthanide coordination, with binding constants up to 10^{28} [1]. The high stability of the [Ln(DOTA)] complexes makes them suitable to vehiculate lanthanide ions in living tissues [2, 3] In particular, [Gd(DOTA)] is largely used as contrast agent in biomedical MRI (DOTAREM^{TM}).

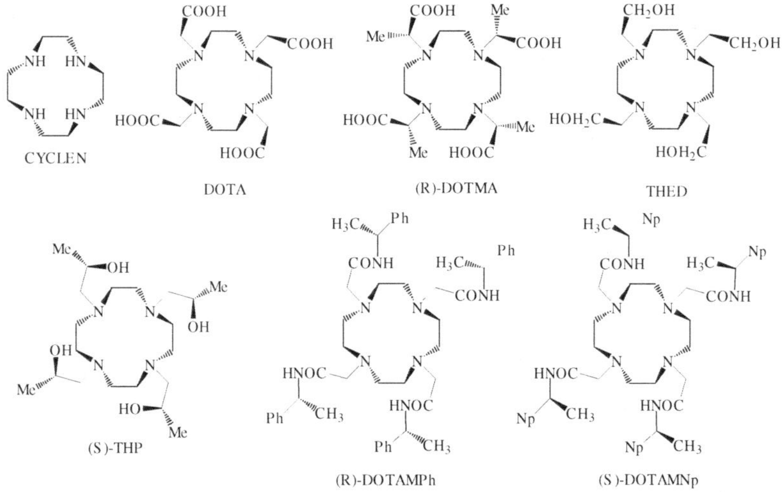

Figure 5.1. DOTA-like ligands discussed in Chapter 5.

In the last years a large interest has been developed in the study of the relation among the structure, the solution dynamics of the complex, and the physico-chemical properties of the lanthanide [3]; in order to arrive to a deeper knowledge of such systems for their growing applications in catalysis, biology and medicine.

This stimulates our interest in investigating in details two of these complexes in solution: the $[Yb((S)\text{-THP})]^{3+}$ (**5**) ((*S*)-THP is 1,4,7,10-tetrakis-((*S*)-2-hydroxypropyl)-1,4,7,10-tetraazacyclododecane) and the $[Yb(THED)]^{3+}$ (**6**) (THED is 1,4,7,10-tetrakis-(2-hydroxyethyl)-1,4,7,10-tetraazacyclododecane), taking advantage of the peculiar spectroscopic properties of ytterbium (see Chapter 2 and 3). The interest for the THP and THED complexes arises from their promising applications as catalysts and contrast agents.

The $[La((S)\text{-THP})]^{3+}$ was demonstrated able to give ion pairs with negatively charged complexes as $[Tm(DOTA)]^-$, $[Tm(DTPA)]^-$, $[Tm(TTHA)]^-$ [4]. More recently, $[Gd((S)\text{-THP})]$ was suggested as potential MRI contrast agent in view of its very fast bound-water exchange [5]. The [Ln(THED)] complexes were found to be effective in promoting the phosphate ester hydrolysis and the transesterification of RNA [6]. In principle, it is possible to realize system for the sequence-selective RNA cleavage by linking these complexes to designed anti-sense RNA oligonucleotides [7]. The mechanism proposed for the phosphate transesterification (Figure 5.2), [8, 9, 10] invokes the direct lanthanide coordination of the phosphate ester and the assistance of a deprotonated side-arm hydroxyl group.

For a deeper understanding of the behavior of such systems, a detailed structural study is required.

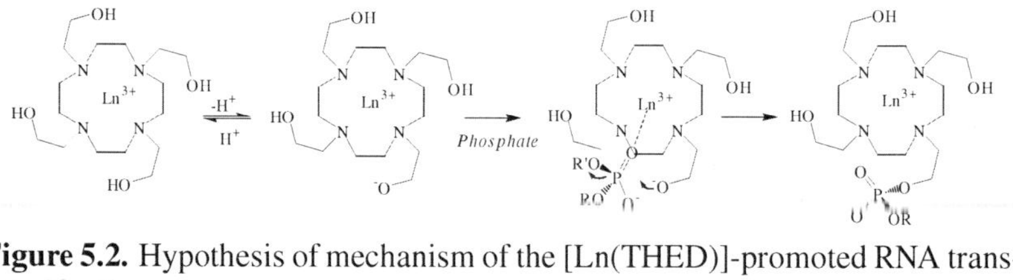

Figure 5.2. Hypothesis of mechanism of the [Ln(THED)]-promoted RNA transesterification.

Indeed, if the structure of the THED complexes with alkali metals [11] is already known, and the solution conformational equilibria of THP and THED complexes with alkali, [12] alkali earth, [13] and Pb(II) ions, [14] were also investigated, [15] no crystallographic structure for [Ln(THED)] and for the homochiral [Ln((*S*)-THP)] complexes are available at the moment. Only two crystal structures of $[Eu((R,S,R,S)\text{-THP})(H_2O)]^{3+}$ and $[Eu((R,R,R,S)\text{-THP})(H_2O)]^{3+}$ heterochiral species have been deposited up to now [16].

5.1. Conformational equilibria in Ln-DOTA-like systems

The lanthanide complexes of DOTA-likes systems have generally similar arrangements and undergo typical conformational equilibria; in order to introduce these aspects, we describe here the ligand arrangement and conformational equilibria observed in [Ln(DOTA)] complexes.

In [Ln(DOTA)] the ligand binds the metal through the four nitrogen atoms of the cyclen ring and the four oxygen atoms of the side arms. The donors arrange in two parallel square faces (Figure 5.3A) tilted by a skew angle ϕ (Figure 5.3B). It descends that the coordination polyhedron is a distorted square-base prism that becomes a regular prism for $\phi = 0°$, and a square antiprism for $\phi = 45°$. Positive values of ϕ are related to Δ helicity of the coordination polyhedron and negative values to the opposite Λ helicity.

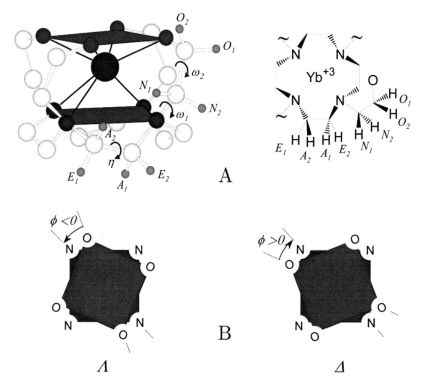

Figure 5.3. A) Ligand arrangement in DOTA-like complexes. The assignment of the hydrogen atoms (green) is reported for the general case of the THED complex, and may be extended to other ligands (DOTA, THP, ...) for their analogue molecular parts. The side-arm protons O_1 and N_1 are assigned to the pro-R positions, while O_2 and N_2 to the pro-S positions. The nitrogen atoms are blue coloured and the oxygen atoms are red coloured. B) Schematic view of the Δ or Λ arrangement of the N- and O- square planes.

The coordination of the metal severely limits the flexibility of the ligand cage to two main dynamics: **side arms rotation**, that displaces the side arms between the Δ and Λ helicity, and the **ring inversion**, which exchanges the ring arrangement between the δ and λ conformations (Figure 5.4). The ring inversion arises from the rotation around the C$-$C bonds of the cyclen and exchanges the positions of the equatorial and axial protons with the correspondence: $A_1 \rightleftharpoons E_1$ and $A_2 \rightleftharpoons E_2$, (see Figure 5.4). The metal coordination imposes the same arrangement to the four N$-$C$-$C$-$N units, so the ring can assume only the two $(\delta,\delta,\delta,\delta)$, and $(\lambda,\lambda,\lambda,\lambda)$ geometries.

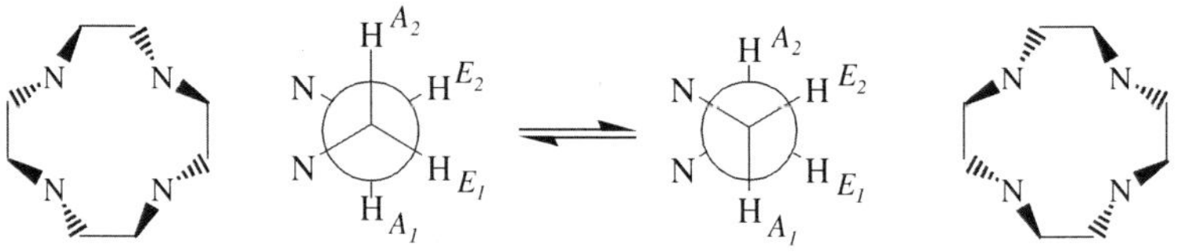

Figure 5.4. The δ and λ ring conformations in [Ln(DOTA-like)] complexes.

The combination of side arm helicity (Δ or Λ) and ring conformations $((\delta,\delta,\delta,\delta)$ or $(\lambda,\lambda,\lambda,\lambda))$ generates four stereoisomeric conformers that in [Ln(DOTA)] are pairwise enantiomers.[1] In the crystallographic structures of DOTA-like complexes, the $\Delta(\lambda,\lambda,\lambda,\lambda)$ and $\Lambda(\delta,\delta,\delta,\delta)$ conformations have $|\phi| \cong 30°$: close to a square-antiprism (*SA*) (this is the case of [Ln(DOTA)], (Ln $=$ Eu [17], Gd [18], Lu [19]) and [Eu(DOTAMPh)] [20]). At variance, the $\Delta(\delta,\delta,\delta,\delta)$ and $\Lambda(\lambda,\lambda,\lambda,\lambda)$ conformations have $|\phi| \cong 15°$ namely twisted square-antiprism (*TSA*) ([La(DOTA)] [21], [La(DOTAM)] [22], [Eu(THP)] [23], [Y(DOTPBz$_4$)] [24], and [Na(DOTAMNp)] [25]).[2]

The NMR study along the lanthanide series, demonstrates that often *TSA* and *SA* conformers are both present in solution, and interconvert through the rotation of the side arm bonds ($\Lambda \rightleftharpoons \Delta$), and ring inversion $((\delta,\delta,\delta,\delta) \rightleftharpoons (\lambda,\lambda,\lambda,\lambda))$ [1, 26]. At high magnetic field these equilibria are slow, on the NMR time scale, and it is possible to resolve all the diasteroisomeric conformations. In particular, for the case of the [Ln(DOTA)], the NMR spectra show two sets of six proton resonances, corresponding to the diasterotopic protons in the $\Delta(\delta,\delta,\delta,\delta)$ and $\Lambda(\lambda,\lambda,\lambda,\lambda)$ conformers (first set) and in the $\Delta(\lambda,\lambda,\lambda,\lambda)$ and $\Lambda(\delta,\delta,\delta,\delta)$ conformations (second set). The slow exchange, on the NMR time scale,

---

[1] The enantiomeric pairs are $(\Delta(\delta,\delta,\delta,\delta); \Lambda(\lambda,\lambda,\lambda,\lambda))$ and $(\Delta(\lambda,\lambda,\lambda,\lambda); \Lambda(\delta,\delta,\delta,\delta))$.

[2] **DOTPBz$_4$** is 1,4,7,10-tetraazacyclododecane-1,4,7,10-tetrakis (methylenebenzylphosphinic acid).

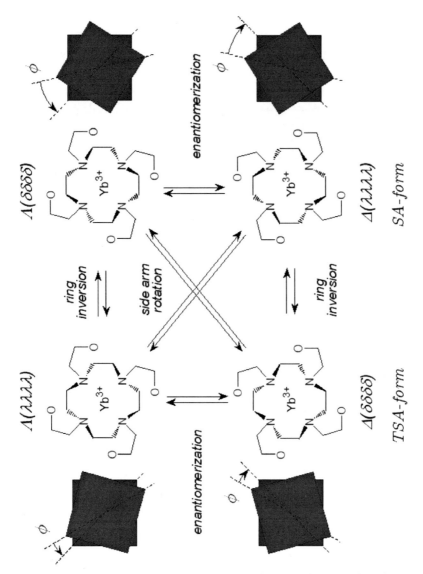

Figure 5.5. Conformational solution equilibria in [Ln(DOTA)] and analogues in solution. Blue and red represent the nitrogen and the oxygen square faces of the coordination polyhedron, respectively.

make possible to resolve the network of conformational equilibria (reported in Figure 5.5) through 2D EXSY experiments [26, 27].[3]

---

[3] The EXSY experiment well evidences the exchanges between the axial and the equatorial ring protons ($A_1 \rightleftharpoons E_1$ and $A_2 \rightleftharpoons E_2$) and the analogous exchange in the methylene protons of the side arms upon side arm rotation.

These dynamics are often flanked by the hydratation/dehydratation equilibria involving water coordination along the axial position [28, 29, 30]. Generally, the affinity for water is different for the TSA and SA conformers, because of the different accessibility of the axial positions. For example, for the $SA \rightleftharpoons TSA$ equilibrium in [Yb(DOTA)], the variable-temperature NMR measures indicate $\Delta H_0 = (17.5 \pm 0.7)$ kJ/mol and $\Delta S_0 = (43 \pm 2)$ kJ/mol. the positive action entropy is attributed to the concomitant dissociation of the coordinated water on going from the major form (SA) to the minor form (TSA), (this is supported also by variable-pressure NMR measures ($\Delta V_0 > 0$)) [28].

In Ln complexes of DOTA-like ligands incorporating stereogenic centers, the pairs $\Delta(\delta,\delta,\delta,\delta)$, $\Lambda(\lambda,\lambda,\lambda,\lambda)$, and $\Delta(\lambda,\lambda,\lambda,\lambda)$, $\Lambda(\delta,\delta,\delta,\delta)$ are no longer enantiomers, and four different diastereoisomeric conformations are obtained.

The chirality of the ligand introduces a further energetic discrimination: in [Yb((R)-DOTMA)] only the $\Lambda(\delta,\delta,\delta,\delta)$ and $\Lambda(\lambda,\lambda,\lambda,\lambda)$ conformations survive, and the TSA species is largely prevalent (major form) [31]. In other complexes as [Yb((R)-DOTAMPh)] and [Yb((S)-DOTAMNp)], only the SA ($\Lambda(\delta,\delta,\delta,\delta)$) conformer is observed [32].

5.2. Synthesis of the (S)-THP and THED complexes

The (S)-THP and THED ligands differ for one methyl group in the side arm (Figure 5.1), which makes (S)-THP a chiral ligand. Both ligands are easily derived from cyclen (Figure 5.6) upon exhaustive addition of ethylene oxide (THED) or propylene oxide (THP).

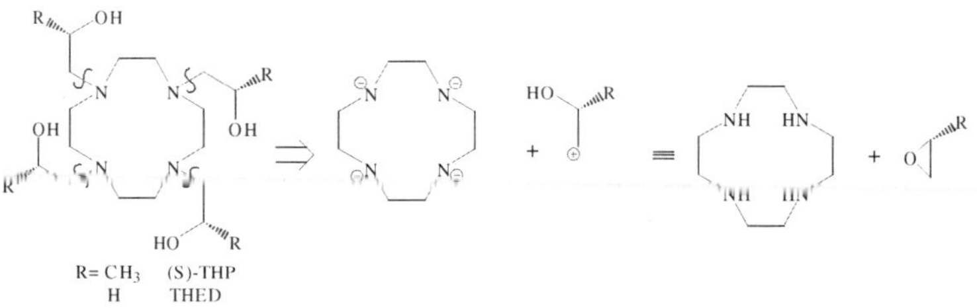

Figure 5.6. Retro-synthetic scheme for THED (R = H) and (S)-THP (R = CH$_3$) ligands.

5.2.1. Synthesis of [Yb((S)-THP)]

Commercially available cyclen (Sigma-Aldrich) was used in the synthesis of (S)-THP (Figure 5.7) [33]. In the first step, the nucleophilic nitrogen atoms of cyclen give addition on the less substituted position of the propylene oxide, with retention of chirality. Using racemic propylene oxide a mixture of six stereoisomers is obtained (two chiral pairs: (R,R,R,R), (S,S,S,S), and (R,R,R,S), (S,S,S,R); and two *meso* forms: (R,R,S,S) and (R,S,R,S)) [16]. To obtain the homochiral (S)-THP enantiopure (S)-propylene oxide must be used (Figure 5.7).

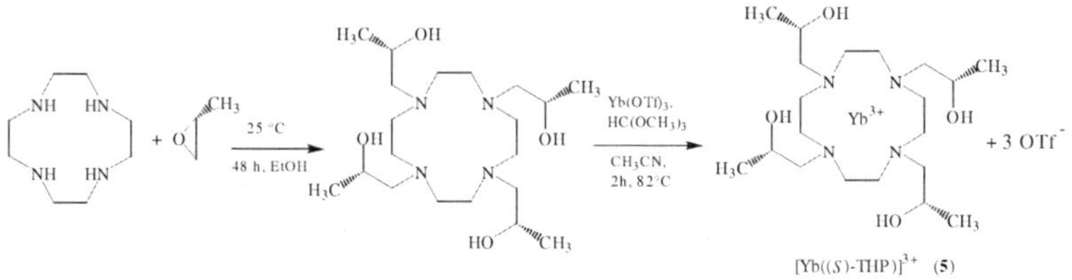

Figure 5.7. Synthesis of **5**.

The reaction was followed through ESI MS spectrometry until the complete disappearance of the mass peaks of the cyclen and of all the partially substituted species. The ligand was purified through crystallization from dry acetonitrile by addition of n-hexane.

The synthesis of the ytterbium complex was conducted under strictly anhydrous conditions to avoid the formation of the lanthanide hydroxide [34]. For this reason, dry Yb(OTf)$_3$ and (S)-THP where stirred in acetonitrile in presence of ethyl ortoformiate as drying agent (Figure 5.7). The complex was purified through crystallization by addition of n-hexane.

5.2.2. Synthesis of [Yb(THED)]

The THED ligand was synthesized from cyclen with exhaustive addition of ethylene oxide [2]. The procedure was similar to that used for (S)-THP (Figure 5.8): ethylene oxide was added as a liquid at $-18\ °C$ in N$_2$ atmosphere, than the reaction bath gradually reached $0\ °C$ and was kept at this temperature for 48 h.

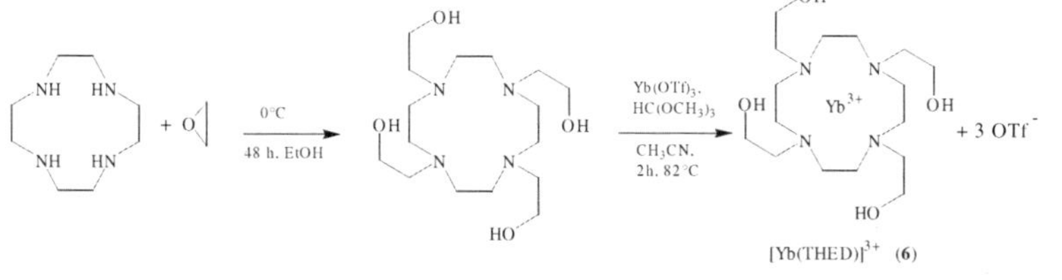

Figure 5.8. Synthesis of **6**.

The reaction was followed with ESI MS to completeness. The ligand was purified in the same manner as (S)-THP. The Yb complex was prepared and purified with a similar procedure as for **5** (Figure 5.8). Also for **6**, anhydrous conditions of synthesis were kept to avoid the formation of ytterbium hydroxide.

5.3. Solution study of [Yb((S)-THP)]

5.3.1. The influence of the pH

The properties of **5** were investigated in solution through NMR, absorption and CD NIR spectroscopy. In water, the ^{1}H NMR resonances of **5** are spread by the paramagnetic interaction over a spectral region between 70 and -50 ppm, with a notable dependence on the pH (Figure 5.9).

The ^{1}H NMR spectra, recorded at $0.79 < \text{pH} < 9$ display *at least* two forms: an acidic form, stable up to pH = 5, and a basic form, stable above pH = 8 (see [35] for details). For $5 < \text{pH} < 8$, the lines of the acidic form progressively broaden and move towards the basic form, in a condition of fast exchange on the NMR time scale (Figure 5.9B). Only 8 resonances are observed, [4] that indicates a stable C_4 symmetry, in analogy with the other LnDOTA-like systems (Section 5.1) [1, 26]. Only one conformer of **5** is present in solution, this also agrees with the ^{1}H NMR spectra of [La((S)-THP)], [Eu((S)-THP)], and [Lu((S)-THP)] [16].

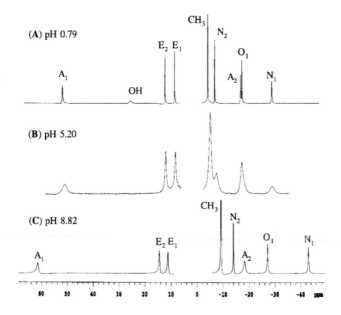

Figure 5.9. ^{1}H NMR spectra of **5** at different pH: **A**) pH = 0.79; **B**) pH = 5.20; **C**) pH = 8.82 (71 mM, solvent = H$_2$O, 600 MHz of proton Larmor frequency, T = 298K). It is noticeable the increased linewidth in the spectrum **B**) due to the proton exchange (the spectra are reported with different vertical scales).

---

[4] In the acidic form only, a broad peak was observed at around 24 ppm, which disappeared in D$_2$O and, consequently, was assigned to the hydroxyl protons.

The proton exchange produces parallel and evident changes in both NIR and NMR spectra. The NIR CD titration well evidences that *only* two chiral species, acidic and basic forms, are involved in the proton exchange (Figure 5.10). The basic species begins to form at pH > 5.5, and at pH > 7.5 it is the only species in solution. This trend is easily followed by plotting the intensity of the NIR CD band at 968 nm, characteristic of the acidic species (Figure 5.11).

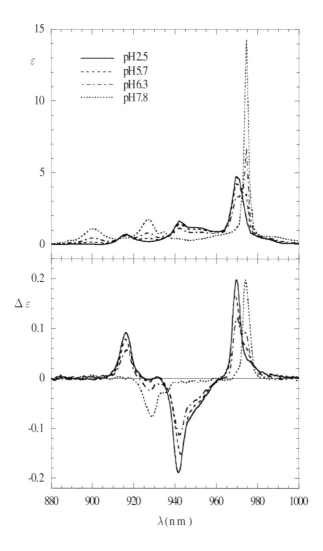

Figure 5.10. NIR spectra of solution of **5** at different pH: **A**) Absortpion spectra; **B**) CD spectra (27 mM, solvent = H_2O, T = 298 K).

The analysis of this plot provides the equilibrium constant for the proton exchange ([5.1], $pK = (6.4 \pm 0.1)$, see [35] for details). This value agrees with that reported for the [Lu((S)-THP)] ($pK = (6.4 \pm 0.1)$) and determined through potentiometric methods [8]. The reported pK for the proton dissociation progressively decreases going from the La to the Lu complexes ([La((S)-THP)] $pK = (8.40 \pm 0.05)$, [Eu((S)-THP)] $pK = (7.8 \pm 0.1)$, [Yb((S)-THP)] $pK = (6.4 \pm 0.1)$, [Lu((S)-THP)][5] $pK_1 = (6.4 \pm 0.1)$, $pK_2 = (9.3 \pm 0.1)$): this trend agrees with the trend of acidity of the lanthanide aquo-ions [36], as expected from the lanthanide contraction (see Chapter 1). In the following (Section 5.3.4), it will be shown that the axial coordination of a molecule of solvent can be excluded for **5**, thus the proton dissociation can derive only from hydroxyl groups, whose acidity is enhanced by the metal coordination (equilibrium [5.1]).

$$[Yb((S)\text{-THP}[-n\text{H}])]^{(3-n)+} \rightleftharpoons [Yb((S)\text{-THP}[-(n+1)\text{H}])]^{(2-n)+} + \text{H}^+$$
$$[5.1]$$

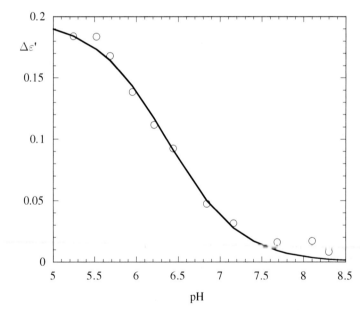

Figure 5.11. Intensity of the CD spectra of **5** at 968 nm as function of pH (27 mM, T = 298 K). The intensity is reported in $\Delta\varepsilon'$ units, which is normalized with respect to the total concentration of complex. The curve fitting is calculated for $pK = (6.4 \pm 0.1)$.

---

[5] Two proton dissociations are observed for the Lu complex: the first (pK_1) is the one comparable with the other complexes.

It is possible that **5** is already deprotonated at pH < 2 ($n > 0$ in (5.1)): conductivity measurements exclude this in the case of [Eu(THP)] [8], but the same may not be true for **5**. In the absence of further evidences we express [5.1] in the most general way, and we still prefer to use the generic terms "acidic form" and "basic form" to address the more and the less protonated species, respectively.

5.3.2. The solution conformation

The ^{1}H NMR spectrum of **5** indicates that only one conformer is present in solution: compared with the achiral **6**, for which four solution conformers are observed (Section 5.4.1), it is evident how the chiral centers in the side arms induce a stereoselection to the whole complex preventing the dynamics reported in Figure 5.5. A similar effect of induction of chirality from the chiral center to the whole ligand cage is observed in the crystal structures of $[Eu((R,S,R,S)\text{-THP})(H_2O)]^{3+}$ and $[Eu((R,R,R,S)\text{-THP})(H_2O)]^{3+}$, where the inversion of chirality of a single side arm is sufficient to switch the complex helicity from Δ (for $[Eu((R,S,R,S)\text{-THP})(H_2O)]^{3+}$) to Λ for $[Eu((R,R,R,S)\text{-THP})(H_2O)]^{3+}$ [16].

In **5** the helicity of the ligand cage was univocally assigned through the structural calculation with the paramagnetic constraints (see Section 5.3.3 and [35]). Indeed, the diastereoisomeric Δ and Λ arrangements have different geometrical factors (especially for the methyl protons) and consequently, different pseudocontact shifts (equation (2.25)). When the experimental pseudocontact shifts are used to optimize the solution structure, only the Δ arrangement produced a reliable structure in good agreement with the NMR data. This arrangement also agrees with that one expected from the crystallographic structure of $[Eu((R,R,R,S)\text{-THP})(H_2O)]^{3+}$ [16].

On the contrary, the analysis of the pseudocontact shifts does not provides any information about the ring conformation $((\delta,\delta,\delta,\delta)$ or $(\lambda,\lambda,\lambda,\lambda))$ and consequently, about the *TSA* or *SA* arrangements. Indeed, it was demonstrated [31] that exchanging the $(\delta,\delta,\delta,\delta)$ and $(\lambda,\lambda,\lambda,\lambda)$ ring conformations the equatorial and axial protons exchange among positions that have the same geometrical factors, and that are scarcely influenced by the side arm orientation (Figure 5.4). Consequently, the experimental pseudocontact shifts agree equally well with both the $\Delta(\delta,\delta,\delta,\delta)$ (*TSA*) or the $\Delta(\lambda,\lambda,\lambda,\lambda)$ (*SA*) geometries. The only way to distinguish between these conformations with NMR, is observing NOE effects between ring and side arm protons, because the *inter*-nuclear distances among these nuclei change in the two arrangements [31]. Unfortunately, any attempt

to measure steady-state NOE effects in **5** gave no positive results, probably in view of its short proton longitudinal relaxation ($T_1 < 10$ ms for all
the ring protons).

In alternative to the NMR methods, we assigned the ring conformation
of **5** using the NIR CD spectrum. NIR CD is very sensitive to the arrangement of the coordination polyhedron, and may be used to compare complexes with similar structures (see Chapter 3) [37]. This is the case, for
example, of [Yb((*S*)-DOTAMNp)] and [Yb((*R*)-DOTAMPh)] that have
both the *SA* arrangement and very similar NIR CD spectra [32], while
the spectrum of [Yb((*R*)-DOTMA)] [38] (which is *TSA*)[6] is completely
different. Noteworthy, the NIR CD spectrum [Yb((*R*)-DOTMA)] is very
similar to that of the acidic form of **5** (Figure 5.12) [35]: in particular,
the three main transitions at wavelengths (970, 941, and 916 nm in **5**)
conserve in both the spectra a sign alternation: $(+, -, +)$ in **5**, $(-, +, -)$
in [Yb((*R*)-DOTMA)].

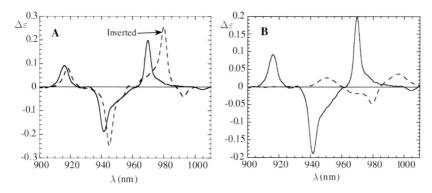

Figure 5.12. A) NIR CD spectra of [Yb((*R*)-DOTMA)] (dashed line) and **5**
(acidic form, continuous line). To better compare these spectra, the spectrum
of [Yb((*R*)-DOTMA)] is inverted. B) NIR CD spectra of [Yb((*R*)-DOTAMPh)]
(dashed line) and **5** (acidic form, continuous line).

The opposite sign of all the bands of [Yb((*R*)-DOTMA)] is related to the
opposite side harm helicity: Λ in [Yb((*R*)-DOTMA)] and Δ in **5**. These
analogies indicate that both complexes have a *TSA* arrangement that is
$\Lambda(\lambda,\lambda,\lambda,\lambda)$ in [Yb((*R*)-DOTMA)] and $\Delta(\delta,\delta,\delta,\delta)$ in **5**.

We do not extend this analysis to the NIR CD spectrum of the basic
form of **5**, which is significantly different from that of the acidic form. It
should be mentioned that the basic form *is not* C_4 symmetric, because of
the lack of one proton: this difference may be important because it de-

---

[6] The NIR CD spectrum is refereed to the major form, which is *TSA*.

stroys the symmetry in the charge distribution of the donors, and consequently in the crystal field. The *intra*-molecular proton exchange, which averages the NMR resonances to an *effective* C_4 symmetry, can not "average" the electronic transition, which operate on a much shorter time scale. As a result, the NIR spectrum of the basic form evidences differences related to a different crystal field (also due to the increased negative charge), and to the molecular geometry that has different symmetry with respect to the acidic form.

5.3.3. The solution structure

The solution structure was determined through the program PERSEUS (Section 2.4) using the proton pseudocontact shifts of the acidic form as structural constraint.[7] The shifts of the basic form and the ^{13}C NMR shifts were not used: the first ones suffer from the intra-molecular proton-exchange, while the second ones have a considerable contact contribution as they are only two bonds away from Yb.

The ^{1}H NMR assignment was obtained by comparing the 1D spectrum with the one of [Yb(DOTA)], on the basis of the integrals, and with the help of a ^{1}H−^{13}C HMQC correlation spectrum (see [35] for details). However, in such a way the stereospecific assignment of the N_1 and N_2 protons remains undetermined between the resonances at −6.6 and −28.6 ppm. This problem was resolved within the structural calculation: two structures were optimized from the two possible assignments, and the one which did not give a reliable fitting was excluded.

The starting model was built up using a MM2 force field [39]. The experimental data (32 proton δ^{pc})[8] allowed PERSEUS to optimize the anisotropy of the magnetic susceptibility tensor (3 parameters, as the tensor is axial (Chapter 2)), the Yb position (3 parameters), and the side arm rotation around the C−N−C−C (ω_1), and N−C−C−O (ω_2) dihedral (2 parameters) (Figure 5.3A). The ring dihedral N−C−C−N (η) shows minimal variation in the [Eu(THP)] crystal structures [16], so this conformation was not changed, reducing the fitting variables.[9]

---

[7] The pseudocontact shifts were calculated using the shifts of [Lu((S)-THP)] as diamagnetic reference.

[8] The calculation was performed on the whole molecule; consequently, all the equivalent protons (es. the four A_1, the four E_2 protons...) were assigned to the same pseudocontact shift. For this reason, the 8 observed resonances correspond to 32 structural restraints.

[9] The dihedral N−C−C−N η was set in the starting structure to $\eta = 55°$, the same values found in the crystal structures of [Eu(THP)] [16]. Anyway, to confirm that this value is also reliable in solution, several solution structures were optimized using starting models in which $45° < \eta < 70°$. These calculations showed that the structures with $\eta = (55 \pm 2)°$ have effectively the best agreement factors R, thus this is definitely accepted for the solution conformation of this complex.

Table 5.1. Optimized geometrical and fitting parameters for the acidic-**5** solution structure. See Figure 5.3 for a definition of the angles η, ω_1, ω_2, ϕ.

Optimized Parameters [Yb((S)-THP)]	
Ring dihedral N−C−C−N η	55° (δ)
Arm dihedral C−N−C−C ω_1	$(69 \pm 3)°$
Arm dihedral N−C−C−O ω_2	$(48 \pm 1)°$
Yb−O	(2.56 ± 0.03) Å
Yb−N	(2.56 ± 0.05) Å
ϕ	$(15.1 \pm 0.2)°$ (Δ)
$\mathcal{D}$	(1550 ± 30) ppm Å³
$R (\delta^{pc})$	5.4%

The two optimized structures show opposite Δ/Λ orientations of the side arm, but the structure with Λ helicity must be discarded because the hydroxyl group is more than 3.6 Å distant from the metal: too far to be coordinated. On the contrary, in the Δ structure the Yb−O distances are 2.6 Å, in agreement with the crystallographic values for eight-coordinated Yb complexes (Chapter 1) [3]. In Figure 5.9 the correct ¹H assignment is reported.

The concomitant use of this structural calculation and the NIR CD analysis (Section 5.3.2) resolved the ambiguity about the ring conformation and assigns to $\Delta(\delta,\delta,\delta,\delta)$ the solution conformation of **5**. In Table 5.1 the optimized parameters relative to this structure are reported: the good agreement factor with the experimental data ($R = 5.4\%$, see Section 2.4) supports the quality of the solution geometry (Figure 5.13).

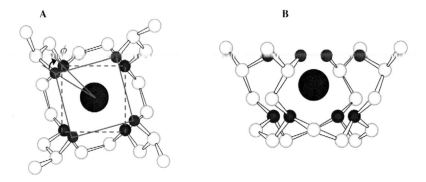

Figure 5.13. Optimized solution structure of the acidic form of **5**. A) View from the top; B) view from a side. The oxygen and the nitrogen atoms are red and blue, respectively. The hydrogen atoms are omitted for clarity.

It should be noticed that, in this structure, the O$-$Yb$-$O bite angle[10] is only 119°, smaller than that of the crystal structure of the [Eu(THP)] complexes (136°). Chin *et al.* suggested that in LnDOTA-like systems, bite angle larger than 130° are typical for axially coordinated complexes (as [Eu(THP)]) [16]: on this basis, **5** is expected to be not-coordinated to water. This hypothesis is fully confirmed in Section 5.3.4 on the basis of the analysis of the NMR spectra in different solvents.

The good agreement factor R between experimental and calculated δ^{pc} indicates that this structure is a good model of the solution geometry. A further improvement of the accuracy of this structure should require the evaluation of contact shift contributions, which are neglected in this approach. In Yb-complexes these contributions are comparably small: Forsberg *et al.* estimated that in analogous [Yb(DOTA-like)] complexes, the contact terms are $< 5\%$ of the corresponding pseudocontact term: only the equatorial ring protons (E_1 and E_2), may have contact contributions up to 30% in view of their arrangement *anti* to metal (Section 2.1.1) [40]. We must consider that an error on a single constraint is largely compensated by the overdetermination of the problem, and that the r^{-3} dependence of δ^{pc} implies that deviations within $\pm10\%$ in the shifts may reflect in error less than ±0.1 Å in the calculated distances, at least for the geometries under the present investigation.

These reasons convinced us that the error introduced in neglecting the contact term is limited and acceptable for the purpose of the structural determination presented here.

5.3.4. Solution dynamics in non-aqueous solvents

The complex **5** was also investigated through NMR in methanol (CD$_3$OD), acetonitrile (CD$_3$CN) and DMSO as the solvents.

In CD$_3$OD, the ^{1}H NMR spectrum shows both the acidic and basic species, that are in slow exchange (on the NMR time scale) and may be resolved (Figure 5.14).

This is probably due to a slower proton exchange in CD$_3$OD with respect to water.[11] The ^{1}H shifts are almost the same as observed in H$_2$O, but with larger linewidths because of the chemical exchange (up to 1400 Hz for A_1 protons at 14 T).

Figure 5.14 also reports the spectra of **5** in aprotic solvents (acetonitrile and DMSO): it is apparent that the shifts of the acidic species are the

---

[10] This angle is measured among the oxygen donors in trans with respect to the metal.

[11] Spectra in CD$_3$OD are recorded in the same conditions as those in water (298 K, 600 MHz of ^{1}H Larmor frequency). The pD was not changed after dissolution, so it is possible that the basic species was formed after a partial deprotonation already during the synthesis.

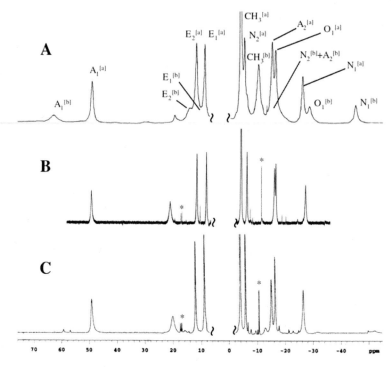

Figure 5.14. ^{1}H NMR spectra of **5** in different solvents: A) CD$_3$OD, B) CD$_3$CN, C) d_6-DMSO (298 K, 14.1 T). In square brackets, the assignments of the acidic form [a] and of the basic form [b] are reported. The peaks with asterisk in CD$_3$CN and d_6-DMSO correspond to small amounts of dimer (see Section 5.3.5).

same in all cases. Two important conclusions can be derived: *a*) **5** has the same solution structure in all these solvents, *b*) the complex is not axially coordinated by the solvent.

Indeed, the axial coordination of the solvent strongly affects the ligand crystal field and the magnetic anisotropy factor $\mathcal{D}$; if the solvent is bound we should observe different spectra [32, 41]. The alternative hypothesis that **5** is capped by a water molecule in all the solvent is quite unreasonable in view of the strong affinity that solvents as DMSO and CD$_3$OD have for lanthanides.

The lack of an axial coordinant is an important difference with respect to [Eu(THP)] [16] and [Gd(THP)] [5]: the crystallographic structure of the former, and ^{17}O NMR evidences for the latter, demonstrate that both are coordinated to water. The smaller Yb ionic radius may be a reason of this fact, but can not exclude that a different ligand conformation influences the solvent accessibility.

5.3.5. Dimerization

The solution of **5** in CD_3CN and d_6-DMSO undergoes a slow evolution during the weeks with the progressive formation of a dimeric species. The process is strongly accelerated by the presence of bases (as Et_3N) or even by a slight excess of free THP ligand. Figure 5.15 reports the ^1H NMR spectrum of the dimer.

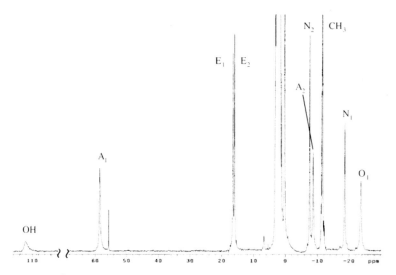

Figure 5.15. ^1H NMR spectrum of the dimeric species of **5** ($[\mathbf{5}]_2$) with the resonance assignment (d_6-DMSO, 298 K, 14.1 T, 40 mM of dimer concentration).

Evidences of the dimerization are obtained though ESI MS: after the addition of Et_3N, it is observed a new peak at $m/z = 1187$ a.m.u. of the monocharged dimeric species $\{[\mathbf{5}-2H^+]_2-H^+\}^+$, which is also confirmed by the peak at $m/z = 575$ a.m.u. of the doubly charged dimeric species $\{[\mathbf{5}-2H^+]_2\}^{2+}$ (Figure 5.16A,B). The absolute charge of these peaks is easily deduced from the isotopic distribution pattern of Yb: in the doubly charged species this pattern features peaks at fractional m/z and cover a width which is only one-half of that of the monocharged peak, as can be noticed comparing the peaks at 576 a.m.u. (Figure 5.16).

The ^1H NMR spectrum of $[\mathbf{5}]_2$ shows only 9 resonances, between 115 and -25 ppm (Figure 5.15), this confirms that this species conserves the main C_4 axis of the monomer and belongs to the D_4 symmetry group. Compared to the spectrum of the monomer, the dimer has much narrower linewidths (<120 Hz for A_1), which are close to the natural linewidths expected for an ytterbium complex. Furthermore, no traces of monomer are present, which means the dimerization process equilibrium is quantitative

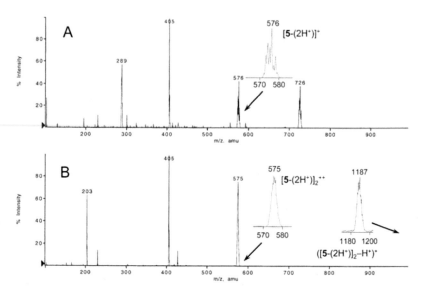

Figure 5.16. ESI MS spectra of **5** in CH_3CN before and after base addition (Et_3N): A) before base addition (monomeric species), B) after base addition (dimeric species). Notice the spreading of the isotopic pattern in the doubly-charged dimeric peak at 576 a.m.u. (B) with respect the mono-charged monomeric peak (A).

and irreversible in these conditions (equilibrium [5.2]).

$$2[Yb((S)\text{-}(THP))]^{3+} \rightleftharpoons [Yb((S)\text{-}(THP-2H^+))]_2^{2+} + 4H^+ \qquad [5.2]$$

The peak at 111 ppm, assigned to the hydroxyl protons, integrates for $1/2$ of the other resonances, that means that each monomeric unit is doubly deprotonated, and the dimer is charged +2, as observed through the ESI MS spectra.

The D_4 symmetry observed in the NMR spectra indicates that the monomers are arranged head-to-head or end-to-end. The simplest model that agrees with all these elements is the head-to-head structure in which the two monomers are joined through four $O \cdots H-O$ hydrogen bonds among the hydroxyl groups. The strength of the hydrogen bonds compensates to the electrostatic repulsion of the charged monomers and gives stability to the dimer. This model, also explains why the dimerization is observed only in CD_3CN and DMSO solutions: in protic solvents (H_2O, CD_3OD) the coupling is hampered by the hydrogen bonds that each monomer establishes with the solvent; furthermore, the dimerization requires that the monomer loses two protons, and this is favored by addition of a base as Et_3N.

The formation of dimers for polydentate macrocyclic complexes was reported for the Co(III) [42], Mn(II) [43], Zn(II) and V(IV) [44] complexes of N, N', N''-tris[(2S)-2-hydroxypropyl]-1,4,7-triazacyclononane, that is a ligand equivalent to THP but derived from the triazacyclononane macrocycle instead of the cyclen. The crystallographic structures of these complexes show an head-to-head coupling analogous to that above proposed for [5]$_2$ with the hydrogen bonds among the hydroxyl groups that hold together the two moieties.

5.3.6. Solution structure of the dimer

The dimeric complex contains two paramagnetic ions, and both affect the shifts and relaxation times of the NMR sensitive nuclei. The ytterbium ions are more than 3 Å away from each other, so their electron-electron coupling may be neglected ($J \ll kT$) and they can be considered isolated [45]. Consequently, the paramagnetic contributions of each Yb ion to a given nucleus are additive and the equations (2.25) and (2.48-50) are rewritten as the sum of the contributions of the two ions (5.1-5)

$$\delta^{pc} = \mathcal{D} \left[\frac{(3 \cos^2 \theta_1 - 1)}{r_1^3} + \frac{(3 \cos^2 \theta_2 - 1)}{r_2^3} \right] \tag{5.1}$$

$$R_1^{dip} = R_2^{dip} = \frac{4}{3} \left(\frac{\mu_0}{4\pi} \right)^2 \gamma_I^2 g_J^2 \mu_B^2 J(J+1) \tau_E \left(\frac{1}{r_1^6} + \frac{1}{r_2^6} \right) \tag{5.2}$$

$$R_1^{Curie} = \frac{6}{7} R_2^{Curie} = \frac{6}{5} \left(\frac{\mu_0}{4\pi} \right)^2 \frac{\gamma_I^2 g_J^2 \mu_B^4 B_0^2 [J(J+1)]^2}{(3kT)^2} \tau_R \left(\frac{1}{r_1^6} + \frac{1}{r_2^6} \right) \tag{5.3}$$

$$R_1^{para} = R_1^{dip} + R_1^{Curie} = k_1 \left(\frac{1}{r_1^6} + \frac{1}{r_2^6} \right) \tag{5.4}$$

$$R_2^{para} = R_2^{dip} + R_2^{Curie} = k_2 \left(\frac{1}{r_1^6} + \frac{1}{r_2^6} \right) \tag{5.5}$$

where r_1 and r_2 are the distances of a nucleus from the two lanthanides, θ_1 and θ_2 are the azimuthal coordinates with respect to the axis adjoining them (Figure 5.18), and k_1 and k_2 are the total paramagnetic relaxation constants for longitudinal and transverse magnetization, respectively. The high symmetry of the dimer imposes that the anisotropy of the magnetic susceptibility $\mathcal{D}$ and the relaxation constants are the same for both the metals. Furthermore, the principal axes of the magnetic anisotropy tensors are collinear with the main C_4 axis.

In Figure 5.17 we report the spatial dependence of the pseudocontact shift around the two metals. The comparison between dimer and monomer δ^{pc} plots (see also Figure 2.4) demonstrates how the nuclei of the dimer experience a different paramagnetic interaction with respect to monomer.

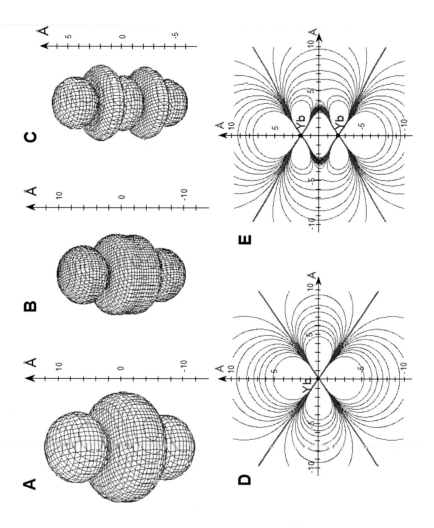

Figure 5.17. Graphical representation of the pseudocontact shift surfaces in the dimer $[5]_2$: A, B, C, reproduce the surfaces with $\delta^{pc} = \pm 5, \pm 10, \pm 50$ ppm, respectively. Red surfaces indicate positive shifts, and blue is for negative shifts. The pictures D and E show the yz cross section of the δ^{pc} surfaces for the single monomeric unit and for the whole dimer, respectively. The black line is calculated for points with $\delta^{pc} = 0$.

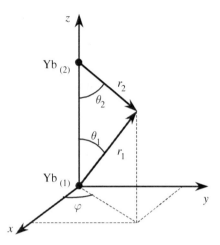

Figure 5.18. Schematic view of the polar coordinate system used in the PERSEUS calculation on the $[5]_2$ complex. The z axis is collinear with the C_4 symmetry axis, the two Yb ions are $Yb_{(1)}$ and $Yb_{(2)}$. In view of the C_4 symmetry, the orientations of x and y axes are irrelevant for the calculation of pseudocontact shift. It should be noticed that the coordinates θ_2 and r_2 are uniquely defined once θ_1, r_1, and the Yb–Yb distance are known.

The pseudocontact shifts and the relaxation rates may still be used in the geometry determination, with a procedure analogue to that used for the monomer. The PERSEUS program was modified introducing the equations (5.1) and (5.4-5) instead of (2.25) and (2.48-50); as constraints, the 32 proton shifts and 24 paramagnetic T_1 were used.[12] The assignment of the NMR spectrum was done through COSY, and [Lu((S)-THP)] [16] was used as the diamagnetic reference.

Two different structures were optimized with PERSEUS: in the first case, the solution structure of each monomeric part was assumed identical to the one of the acidic form described above. The distance between the two moieties (*i.e.* between the two ytterbium ions) was varied (one parameter) together with the magnetic parameters (3 parameters for the magnetic susceptibility tensor, and one parameter for the relaxation constant). The second structure was optimized allowing the structural rearrangement of the monomeric units: in addition to the variables optimized in the first model, PERSEUS optimized also the side arm and the

---

[12] The paramagnetic shift of the hydroxyl group was not used because it is only two bonds length away Yb, then the contact contribution is not negligible. The longitudinal relaxation times of the methyl group and of the A_1 protons are not used because they are very short and suffer from a considerable error.

cyclen ring conformations (rearranged structure). In Table 5.2 we report the most important structural parameters for these two models and their minimized agreement factors R. The rearranged structure (Figure 5.19) agrees much better with the experimental data (see agreement factor in Table 5.2); besides, the first model can not be accepted because it leads to a Yb−Yb distance of only 2.2 Å, that is too small. On the contrary, in the rearranged structure the Yb atoms are 4.38 Å away, which is enough to allow the formation of hydrogen bonds between the hydroxyl groups of the two units (O· · ·H−O around 2.6 Å) [46]. The principal structural changes are the rotation of the side arms helicity that go from Δ (in the monomer) to Λ; minor changes also affect the cyclen ring.

Table 5.2. Optimized geometrical and fitting parameters for $[5]_2$. The second column refers to the dimeric structure calculated without rearranging the ligand cage. In the third column the data refers to the solution structure with a conformational rearrangement of the ligand cage.

Optimized Parameters	Dimer	Rearranged dimer
Ring dihedral N−C−C−N η	$55°$ (δ)	$(60 \pm 3)°$ (δ)
Arm dihedral C−N−C−C ω_1	$(69 \pm 3)°$	$(96 \pm 1)°$
Arm dihedral N−C−C−O ω_2	$(48 \pm 1)°$	$(53 \pm 1)°$
Yb−Yb	(2.2 ± 0.7) Å	(4.38 ± 0.12) Å
Yb−O	(2.56 ± 0.03) Å	(2.46 ± 0.05) Å
Yb−N	(2.56 ± 0.05) Å	(2.59 ± 0.03) Å
ϕ	$(15.1 \pm 0.2)°$ (Δ)	$(-16.6 \pm 0.2)°$ (Λ)
$\mathcal{D}$	(1300 ± 170) ppm Å^3	(1770 ± 20) ppm Å^3
R (δ^{pc})	12%	4.5%
R (ρ^{para})	52%	2.9%

It appears that the dimerization induces the rearrangement of the monomeric moieties in order to allow an easier interaction between the complexes, and to reduce the repulsion of the methyl groups. These are probably the driving forces that invert the side arm helicity to the Λ conformation.

The structural calculation determines the relative distances between the monomeric units, but not their reciprocal twist. Unfortunately, this supramolecular arrangement can not be easily investigated through the paramagnetic NMR analysis, indeed the relative rotation around the C_4 axis of one monomeric moiety with respect to the other does not affect none of the r_1, r_2, θ_1, θ_2, parameters of (5.1) and (5.4-5), and conse-

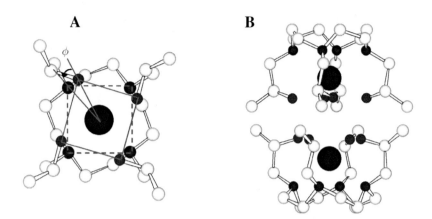

Figure 5.19. Optimized solution structure of the rearranged dimer [**5**]$_2$: A) view from the top of the single monomeric moiety, B) view from a side of the whole dimer. The hydrogen atoms are omitted: carbons are in grey, oxygens are in red, and nitrogens are in blue. Notice the inverted side arm arrangement with respect to the monomeric species (Figure 5.13).

quently it does not produce any effect on the NMR resonances. In the model reported in Figure 5.19 the relative twist of the two monomers is reasonably fixed imposing the O $\cdots$ H$-$O hydrogen bonds close to their typical values [46], but further investigation on this aspect is not performed in this thesis.

5.3.7. NIR CD study of the dimer

The NIR CD spectrum of the dimer shows a completely different profile with respect to the monomer (Figure 5.20). In the dimer the Cotton effects are less intense, with two main transitions at 940 and 975 nm, with sign $+$ and $-$, respectively.

In the absorption spectrum other transitions from 880 to 1020 nm are observed, but they do not give appreciable CD. It should be noticed that the spectra of the dimer are very similar to those one of the basic form of **5** (Figure 5.10), even if they are less intense and the CD-bands have inverted sign.

It is possible that the ligand crystal field terms of the basic form are similar to that of the dimer, because in both the cases the ligand is partially deprotonated; furthermore, the opposite sign of the NIR CD bands in the dimer could be reasonably explained as the helicity of the side arm is inverted passing from Δ, in the monomer, to Λ in the dimer (see Section 5.3.3). Anyway, these hypotheses are only indicative, and the ef-

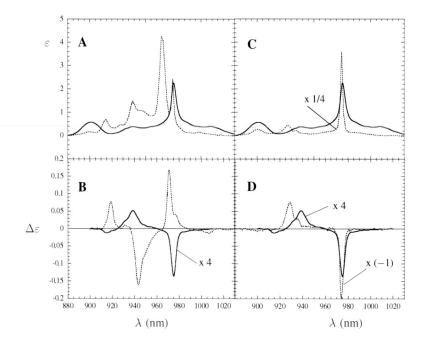

Figure 5.20. Absorption (A) and NIR CD (B) spectra of [**5**]$_2$ in CH$_3$CN (continuous line, c = 40 mM) and of **5** in CH$_3$CN (dotted line, c = 80 mM). In C) and D), the same spectra of [**5**]$_2$ in CH$_3$CN are compared with the spectra of the basic form of **5** in H$_2$O (dotted line) (pH = 7.8). To make the comparison easier, the CD spectra of the dimer in B) and D) are multiplied by 4, the absorption spectrum of the basic **5** (C) is divided by 4, and the CD spectrum of the basic **5** is inverted.

fect of the structural rearrangement of the dimer on the NIR CD spectra should require further investigations.

5.4. Solution study of [Yb(THED)]

5.4.1. Solution investigation in non-aqueous solvents

The [Yb(THED)]$^{3+}$ (**6**) has a very low solubility in water, which does not appreciably increases even varying the pH within the 0–9 range. On the contrary, **6** is well soluble in CD$_3$OD, CD$_3$ CN and DMSO, and the structural studies reported here are referred to these solvents.

The ^{1}H NMR spectrum in acetonitrile shows resonances dispersed from 180 to –80 ppm (Figure 5.21). Two sets of resonances are distinguished and attributed to a major and a minor form. The 2D EXSY experiment (Figure 5.22) demonstrates that these forms are in slow exchange (on the NMR time scale) and corresponds to the *SA* and *TSA* con-

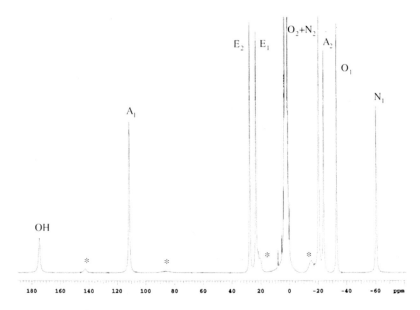

Figure 5.21. ^{1}H NMR spectrum of **6** with resonance assignment (CD$_3$CN, 298 K, 14.1 T). Stared peaks indicate the minor conformer (see in the text).

formations depicted in Figure 5.5, and expected in view of the structural analogy with [Yb(DOTA)].

In the major form 9 different protons are observed,[13] while the resonances of the minor form are broad and partially overlapped with those of the major form, and a complete set of resonances is not observable.

The ^{1}H NMR spectra of **6** in methanol and in DMSO show the peaks of the major form essentially at the same frequencies observed in CD$_3$CN. On the contrary, the resonances of the minor form shift considerably, and sometimes are so broad that are hardly observed. In analogy with **5**, one must conclude that the major form has the same structure in all these solvents and that is not axially coordinated (CN = 8, see Section 5.3.4). On the contrary, it is still possible that the minor form is subject to axial coordination dynamics, involving the solvent, which explains why its spectrum is solvent dependent.

5.4.2. Solution dynamics in CD$_3$CN

The linewidth of the ^{1}H NMR resonances in Figure 5.21 are much larger than those observed for **5** (in **6** the A_1 proton is 375 Hz broad, the cor-

---

[13] The resonance at –20 ppm integrates for two protons.

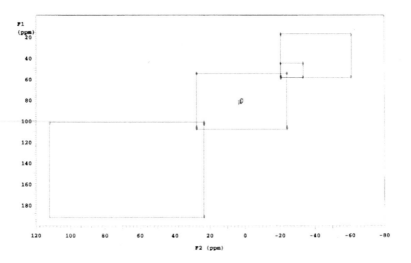

Figure 5.22. EXSY spectrum of **6** in CD₃CN. The boxes indicate the exchanging peaks. Only the major form is visible with this threshold level.

responding resonance in the monomer of **5** is 170 Hz and in the dimer of **5** is <120 Hz), and they are also broader than those expected for the paramagnetic enhanced relaxation (see Section 4.4.1 and [47]). Through a quantitative analysis of the 2D EXSY [48, 49], it was evaluated that the conformational exchange may account for the proton line broadening, but one can not exclude also a minor contribution due to proton exchange among the hydroxylic groups.

The NMR investigation of the conformational exchange was extended to the temperature range within 25 °C and −45 °C (Figure 5.23). Cooling down the sample from 25 °C to 0 °C the linewidths in the ¹H NMR spectra decrease: this confirms that the peak broadening is dominated by exchange processes, indeed, if the line broadening is derived from the paramagnetic effect, the linewidth should increase when lowering the temperature (2.48-50). The combined effect of the line narrowing and the increased spectral width (expected from the T^{-2} trend of the $\mathcal{D}$ factor, Section 2.3.1), gives an improved resolution that allows one to distinguish many of the minor form resonances (Figure 5.23).

Cooling down below 0 °C two distinct effects are observed for the resonances of the major and the minor forms. For the major form, the line narrowing proceeds up to −20 °C: below this temperature the paramagnetic line broadening dominates and the peak linewidths increase. For the minor form, all the resonances progressively split into two new and almost equally intense peaks, with the exception of the −OH protons (140 ppm).

This difference in behaviour between minor and major form can be explained in the hypothesis that the minor form is partially solvent coordinated. At room temperature the solvent exchange dynamics of the minor form is fast (on the NMR timescale), and average the resonances of bound and unbound forms, so only one set of resonances are observed. By lowering temperature below 0 °C this exchange becomes slow on the NMR timescale, and the resonances of bound and unbound forms split. In DOTA-like complexes indeed the water exchange is generally fast ($10^6 - 10^7$ s^{-1}) [5], but it can decrease in complexes able to give hydrogen bonds with the bound water [29]. This probably should be the case of the minor form of **6**.

For the major form no-solvent coordination is expected and the resonances remain not split.

In summary, the [Yb(THED)]$^{3+}$ in solution undergoes two main exchange dynamics: the conformational exchange and the coordinated solvent exchange. The first operate through the ring inversion and side arm rotation dynamics, and exchange between the resonances of the major and minor forms (as depicted in Figure 5.5). In the present conditions (CD$_3$CN, 14.1 T) and at room temperature these equilibria are slow on the NMR time scale, and are responsible for the most of the observed linewidths. By lowering the temperature the conformational exchange is progressively frozen and the linewidth of the major and minor form reduce. Instead, the coordinated solvent exchange acts only on the minor form, and it is fast on the NMR timescale and at room temperature. It becomes slow only below 0 °C, where the resonances of the bound and unbound minor form are split. As the major form is not solvent bound, this process is absent for this species.

Additional dynamics as the proton exchange also should occur involving the hydroxyl protons of major and minor forms as well as traces of water present in the solvent. This may be a reason of the different behaviour observed for the hydroxyl protons of the minor form during the variable temperature study. Anyway, this dynamics affects the spectra much less of those described above and was not further investigated in the present thesis.

5.4.3. Solution structure

The THED and the DOTA ligands are very similar, and differentiate only for the terminal part of the side arms: a –CH$_2$OH group in THED and –COO$^-$ group in DOTA. Similar geometries can be expected also for their lanthanide complexes.

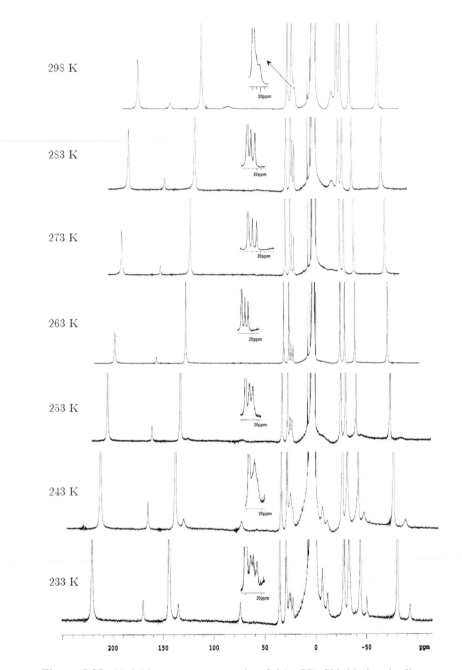

Figure 5.23. Variable-temperature study of **6** in CD$_3$CN. Notice the line narrowing and split of the resonances of the minor form. The expansion evidences these processes for the minor-form resonances around 20 ppm.

Experimental evidences of the structural analogies between [Yb(DOTA)] and **6** are apparent by simply comparing the ^{1}H NMR spectra of the two complexes, showing a correspondence in the peak positions. This correspondence between the 1D spectra, together with the knowledge of the exchanging pairs ($A_1 \rightleftharpoons E_1$, $A_2 \rightleftharpoons E_2$, $N_1 \rightleftharpoons N_2$, $O_1 \rightleftharpoons O_2$) in the 2D EXSY of **6**, made possible to assign the spectrum (Figure 5.21).[14] The ^{13}C shifts were successively assigned through ^{1}H$-^{13}$C HMQC spectra.

A more detailed comparison of the structure of [Yb(DOTA)] and **6**, is performed plotting the ^{1}H pseudocontact shifts of **6** against the corresponding shifts of [Yb(DOTA)] (Figure 5.24).[15] As the δ^{pc}'s are related to the molecular geometry (equation (2.25)), if the two complexes have the same structure, they must have also the same *GF*s ($GF = (3\cos^2\theta - 1)/r^3$), and consequently proportional pseudocontact shifts.

The straight line showed in Figure 5.24 is a good evidence of the structural analogies of the two complexes, but it is not sufficient to demonstrate that they have the *same* structure. Indeed, it was demonstrated that both the *TSA* or *SA* arrangement might corresponds to the same set of *GF* [31]: the plot of Figure 5.24 indicates that the major form of **6** may be *SA*, namely it has the same structure of the major form of [Yb(DOTA)], but it is also possible that **6** have a precise *TSA* conformation with the same *GF*s of [Yb(DOTA)].

Unfortunately, any attempt to observe NOE's among the ring and side arm protons, in order to distinguish between these two conformations [31], was unsuccessful, probably because of the short relaxation times. Furthermore, **6** is achiral, and the method based on the NIR CD correlation (Section 5.3.1) can not be performed. Thus, no sure distinction between *SA* or *TSA* is at the moment possible; the only indications that propend for the *TSA* arrangement is the absence of water axially bound, which in [Yb(DOTA)] [28] and [Yb((R)-DOTMA)] [31] is typical for the *TSA* arrangement.

With the paramagnetic analysis (PERSEUS calculation) we were able to refine the solution structure of **6**, and also to assign the conformation of the CH$_2$OH group that is not present in [Yb(DOTA)]. We decided to optimize only the *TSA* geometry even if, as above noted, the same calculation can be performed for the *SA* arrangement too. The final structure (reported in Figure 5.25) shows the $\Delta(\delta,\delta,\delta,\delta)$ conformation: the geometry of the $\Lambda(\lambda,\lambda,\lambda,\lambda)$ can be simply derived from a mirror im-

---

[14] Only the stereospecific assignment of the O_1/O_2 protons remained undetermined, but this is resolved within the structural calculation, (see in the text).

[15] This comparison can be done for all the protons of **6** except O_1 and O_2, which are absent in [Yb(DOTA)].

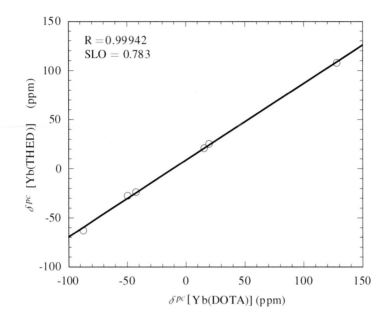

Figure 5.24. Plot of the pseudocontact shifts of **6** (in CD$_3$CN) against the pseudocontact shifts of the corresponding protons in the major form of [Yb(DOTA)] in D$_2$O.

age. The starting model was built using a MM2 force field and imposing the N−C−C−N ring dihedrals to $\eta = 59°$ (δ arrangement), which is the same angle found in the crystallographic [Ln(DOTA)] structures [17, 18, 19]

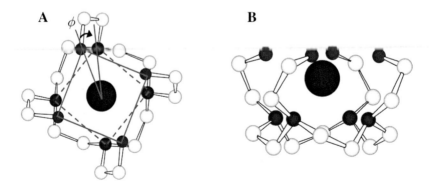

Figure 5.25. Optimized solution structure of **6** (A, view from the top; B, view from a side). The hydrogen atoms are omitted: carbons are in grey, oxygens are in red, and nitrogens are in blue, the ytterbium is in brown.

The stereospecific assignment of the O_1/O_2 protons, which can not be obtained by comparison with the [Yb(DOTA)] shifts, is determined with the structural calculation with a procedure analogue to the one used for the complex **5** (Section 5.3.1). So, two structures were optimized using the two possible O_1/O_2 assignments, and the one that gives a not acceptable structure was discarded. In the structural calculation also the paramagnetic tensor (3 parameters), the complete side arm conformation (2 dihedral angles) and the metal position (3 parameters) were optimised, using the 32 δ^{pc} of the ring and side arm protons. The very good agreement factor ($R = 3.6\%$, Table 5.3) supports the quality of the calculated structure (Figure 5.25).

Table 5.3. Structural and magnetic parameters optimised in the PERSEUS calculation of the solution structure of **6**.

Optimized Parameters [Yb(THED)]	
Ring dihedral N−C−C−N η	$59°$ (δ)
Arm dihedral C−N−C−C ω_1	$(72 \pm 3)°$
Arm dihedral N−C−C−O ω_2	$(30.0 \pm 1.5)°$
Yb−O	(2.44 ± 0.03) Å
Yb−N	(2.64 ± 0.03) Å
ϕ	$(20.8 \pm 0.3)°$ (Δ)
$\mathcal{D}$	(3690 ± 40) ppm Å^3
R (δ^{pc})	3.6%

5.5. Conclusions

NIR CD and NMR are applied here to the study in solution of [Yb((S)-THP)] and [Yb(THED)], two members of the family of the lanthanide DOTA-like complexes. The strict dependence of the paramagnetic interactions on the molecular geometry makes the analysis of the paramagnetic NMR contributions (as pseudocontact shift and relaxation times) a unique and sensitive tool for the structural investigation in solution. This allowed us not only to determine refined structure in solution of such complexes, but also to analyse the conformational dynamics and the structural rearrangement during the dimerization process, which are hardly investigated otherwise. NIR CD spectroscopy completed the structural investigation performed through NMR providing information about the chiral arrangement of the ligand, and making possible to study dynamic events like proton dissociation or solvent coordination equilibria. This is a suitable technique to study complexes that are not endowed with chromophores absorbing in the UV-Vis region.

More in detail, the solution study carried out on [Yb((S)-THP)] (**5**) indicates that it is present in solution as a single conformer; which means that the chiral centers in the side arms freeze both side arm rotation and the ring inversion dynamics (Figure 5.5). The detailed analysis of the paramagnetic NMR contributions, and of the NIR CD spectrum allowed us to assign the ligand conformation to the *TSA* $\Delta(\delta,\delta,\delta,\delta)$ arrangement. In protic solvents (H_2O, methanol) **5** undergoes to a proton exchange, and loses one proton from the hydroxyl group at basic pH. This process is easily followed through NIR CD, which makes possible to determine the $pK = (6.4 \pm 0.1)$, in good agreement with the constants determined through potentiometric methods on analogous [Ln(THP)] complexes. The solution structure of **5** was found with the help of the computer program PERSEUS, and using the pseudocontact shifts of the acidic species as experimental constraints. The NMR spectra recorded in several solvents (H_2O, methanol, acetonitrile, DMSO) indicates that the structure of the complex is conserved and that the solvent is not axially bound to the metal.

In non-protic solvents (acetonitrile, DMSO), and in the presence of small amounts of bases (as Et_3N) **5** dimerizes yielding to a head-to-head structure maintaining the C_4 symmetry axis of the monomer. The process was followed through ESI MS, NMR and NIR CD spectroscopy. The solution structure, determined on the basis of the pseudocontact shifts and the paramagnetic relaxation times, indicates that each monomeric unit undergoes a structural rearrangement upon dimerization, by changing the helicity of the side arms from Δ to Λ, and with minor variations of the ring conformation. The two monomeric units are held together through four hydrogen bonds that involve the hydroxyl groups. The NIR CD spectrum of the dimer radically changes with respect to the monomer, supporting the structural rearrangement of the ligand cage.

The achiral analogous of **5**, the complex [Yb(THED)] (**6**), has a solution behaviour quite different from **5**. The absence of chiral centers allows a larger flexibility of the ligand cage and both side arm rotation and ring inversion are active in solution (acetonitrile), with the formation of a major and a minor form (Figure 5.5). The major form is not solvent coordinated, where the minor form is partially bound to the solvent. NMR spectra recorded from -40 to 25 °C showed that, at low temperature, the solvent exchange for the minor form becomes slow (on the NMR time scale), and the two bound and unbound species can be identified.

The calculation of the solution structure of **6** shows that this complex has a ligand arrangement very similar to that of the *TSA/SA* in

[Yb(DOTA)], but the NMR data alone does not allow us to determine the exact conformations of the major and minor forms.

All these results indicate that the complexes **5** and **6** in solution undergo several equilibria in solution: the knowledge of such processes and the development of methods for their investigation are of primary importance in the applications, especially in catalysis, for which ligand arrangement, conformational equilibria, coordination of the axial position are often determinant in the catalytic mechanism.

References

[1] DESREUX, J. F., *Inorg. Chem.* **1980**, *19*, 1319–1324.

[2] MORROW, J. R.; CHIN, K. A. O., *Inorg. Chem.* **1993**, *32*, 3357–3361.

[3] PARKER, D.; DICKINS, R. S.; PUSCHMANN, H.; CLOSSLAND, C.; HOWARD, J. A. K., *Chem. Rev.* **2002**, *102*, 1977–2010.

[4] CORSI, D. M.; VAN BEKKUM, H.; PETERS, J. A., *Inorg. Chem.* **2000**, 4802–4808.

[5] CORSI, D. M.; ELST, L. V.; MULLER, R. N.; VAN BEKKUM, H.; PETERS, J. A., *Chem. Eur. J.* **2001**, 1383–1389.

[6] MORROW, J. R.; BUTTREY, L. A.; SHELTON, V. M.; BERBACK, K. A., *J. Am. Chem. Soc.* **1992**, *114*, 1903–1905.

[7] (a) TRAWICK, B. N.; DANIHER, A. T.; BASHKIN, J. K., *Chem. Rev.* **1998**, *98*, 939–960. (b) BAKER, B. F.; KHALILI, H.; WEI, N.; MORROW, J. R., *J. Am. Chem. Soc.* **1997**, *119*, 8749–8755 and references therein.

[8] CHIN, K. O. A.; MORROW, J. R., *Inorg. Chem.* **1994**, *33*, 5036–5041.

[9] MORROW, J. R.; AURES, K.; EPSTEIN, D., *J. Chem. Soc., Chem. Commun.* **1995**, 2431–2432.

[10] EPSTEIN, D. M.; CHAPPELL, L. L.; KHALILI, H.; SUPKOWSKI, R. M.; HORROCKS, W. D. J.; MORROW, J. R., *Inorg. Chem.* **2000**, *39*, 2130–2134.

[11] BUØEN, S.; DALE, J.; GROTH, P.; KRANE, J., *J. Chem. Soc., Chem. Commun.* **1982**, 1172–1174.

[12] (a) DHILLON, R. S.; MADBAK, S. E.; CICCONE, F. G.; BUNTINE, M. A.; LINCOLN, S. F.; WAINWRIGHT, K. P., *J. Am. Chem. Soc.* **1997**, *119*, 6126–6134. (b) DHILLON, R.; STEPHENS, A. K. W.; WHITBREAD, S. L.; LINCOLN, S. F.; WAINWRIGHT, K. P., *J. Chem. Soc., Chem. Commun.* **1995**, 97–98. (c) WHITBREAD, S. L.; POLITIS, S.; STEPHENS, A. K. W.; LUCAS, J. B.; DHILLON, R.; LINCOLN, S. F.; WAINWRIGHT, K. P., *J. Chem. Soc., Dalton Trans.* **1996**, 1379–1384.

[13] DHILLON, R.; LINCOLN, S. F.; MADBAK, S.; STEPHENS, A. K. W.; WAINWRIGHT, K. P.; WHITBREAD, S. L., *Inorg. Chem.* **2000**, *39*, 1855–1858.

[14] PITTET, P.-A.; LAURENCE, G. S.; LINCOLN, S. F.; TURONEK, M. L.; WAINWRIGHT, K. P., *J. Chem. Soc., Chem. Commun.* **1991**, 1205–1206.

[15] PITTET, P.-A.; FRÜH, D.; TISSIÈRES, V.; BÜNZLI, J.-C. G., *J. Chem. Soc., Dalton Trans.* **1997**, 895–900.

[16] CHIN, K. O. A.; MORROW, J. R.; LAKE, C. H.; CHURCHILL, M. R., *Inorg. Chem.* **1994**, *33*, 656–664.

[17] **EuDOTA**: Spirlet, M. R.; Rebizant J.; Desreux, J. F.; Loncin, M. F., *Inorg. Chem.* **1984**, *23*, 359–363.

[18] **GdDOTA**: Dubost, J.-P.; Legar, M.; Meyer, D.; Schaefer, M., *C. R. Acad. Sci. Paris, Ser. 2* **1991**, *312*, 349–354.

[19] **LuDOTA**: Aime, S.; Barge, A.; Botta, M.; Fasano, M.; Ayala, J. D.; Bombieri, G., *Inorg. Chim. Acta* **1993**, *246*, 423–429.

[20] **EuDOTAMPh**: Dickins, R. S.; Howard, J. A. K.; Lehmann, C. W.; Moloney, J.; Parker, D.; Peacock, R., *Angew. Chem. Int. Ed. Engl.* **1997**, *36*, 521–523.

[21] **LaDOTA**: Aime, S.; Barge, A.; Benetollo, F.; Bombieri, G.; Botta, M.; Uggeri, F., *Inorg. Chem.* **1997**, *36*, 4287–4289.

[22] **LaDOTAM**: Morrow, J. R.; Amin, S.; Lake, C. H.; Churchill, M. R., *Inorg. Chem.* **1993**, *32*, 4566–4572.

[23] **EuTHP**: Chin, K. O. A.; Morrow, J. R.; Lake, C. H.; Churchill, M. R., *Inorg. Chem.* **1994**, *33*, 656–664.

[24] **YDOTPBz₄**: Aime, S.; Batsanov, S.; Botta, M.; Howard, J. A. K.; Parker, D.; Senanayake, K.; Williams, G., *Inorg. Chem.* **1994**, *33*, 4696–4706.

[25] **NaDOTAMNp**: Dickins, R. S.; Howard, J. A. K.; Moloney, J.; Parker, D.; Peacock, R.; Siligardi, G., *J. Chem. Soc., Chem. Commun.* **1997**, 1747–1748.

[26] Aime, S.; Botta, M.; Ermondi, G., *Inorg. Chem.* **1992**, *31*, 4291–4299.

[27] Jacques, V.; Desreux, J. F., *Inorg. Chem.* **1994**, *33*, 4048–4053.

[28] Aime, S.; Botta, M.; Fasano, M.; Marques, M. P. M.; Geraldez, C. F. G. C.; Pubanz, D.; Merbach, A. E., *Inorg. Chem.* **1997**, *36*, 2059–2068.

[29] Aime, S.; Barge, A.; Botta, M.; De Sousa, A. S.; Parker, D., *Angew. Chem. Int. Ed.* **1998**, *37*, 2673–2674.

[30] Aime, S.; Barge, A.; Bruce, J. I.; Botta, M.; Howard, J. A. K.; Moloney, J. M.; Parker, D.; De Sousa, A. S.; Woods, M., *J. Am. Chem. Soc.* **1999**, *121*, 5762–5771.

[31] Di Bari, L.; Pintacuda, G.; Salvadori, P., *Eur. J. Inorg. Chem.* **2000**, 75–82.

[32] Di Bari, L.; Pintacuda, G.; Salvadori, P.; Dickins, R. S.; Parker, D., *J. Am. Chem. Soc.* **2000**, *122*, 9257–9264.

[33] Hancock, R. D.; Shaikjee, M. S.; Dobson, S. M.; Boeyens, J. C. A., *Inorg. Chim. Acta* **1988**, *154*, 229–238.

[34] Kobayashi, S. *Lanthanides: Chemistry and Use in Organic Synthesis, Topics in Organometallic Chemistry,* **1999**, Springer-Verlag: Berlin.

[35] Lelli, M.; Pintacuda, G.; Cuzzola, A.; Di Bari, L., *Chirality* **2005**, *17*, 201–211.

[36] Maretll, A. E.; Smith, R. M., *NIST Standard Reference Database 46 Version 6.0,* **2001**.

[37] Di Bari, L.; Pescitelli, G.; Sherry, A. D.; Woods, M., *Inorg. Chem.* **2005**, *44*, 8391–8398.

[38] Di Bari, L.; Pintacuda, G.; Salvadori, P. *J. Am. Chem. Soc.* **2000**, *122*, 5557–5562.

[39] CS CHEM 3D PRO V.3.5.2, Cambridge Soft Corporation; Cambridge, Massachusetts, 02139 USA.
[40] FORSBERG, J. H.; DELANEY, R. M.; ZHAO, Q.; HARAKAS, G.; CHANDRAN, R., *Inorg. Chem.* **1995**, *34,* 3705–3715.
[41] DICKINS, R. S.; PARKER, D.; BRUCE, J. I.; TOZER, D. J., *Dalton. Trans.* **2003**, 1264–1271.
[42] BELAL, A. A.; FARRUGIA, L. J.; PEACOCK, R. D., ROBB, J., *J. Chem. Soc., Dalton Trans.* **1989**, 931–935.
[43] BELAL, A. A.; CHAUDHURI, P.; FALLIS, I.; FARRUGIA, L. J.; HARTUNG, R.; MACDONAND, N. M.; NUBER, B.; PEACOCK, R. D., WEISS, J.; WIEGHARDT, K., *Inorg. Chem.* **1991**, *30,* 4397–4402.
[44] FALLIS, I.; FARRUGIA, L. J.; MACDONAND, N. M.; PEACOCK, R. D., *Inorg. Chem.* **1993**, *32,* 779–780.
[45] (a) BERTINI, I.; LUCHINAT, C., *Coord. Chem. Rev.* **1996**, *150.* (b) BERTINI, I.; LUCHINAT, C.; PARIGI, G., *Solution NMR of Paramagnetic Molecules,* **2001**, Elsevier: Amsterdam. (c) BERTINI, I.; LUCHINAT, C.; PARIGI, G., *Prog. Nucl. Magn. Reson. Spec.* **2002**, *40,* 249–273.
[46] COTTON F.A.; WILKINSON, G., *Advanced Inorganic Chemistry, Fifth Edition,* **1988**, John Wiley & Sons, Inc.: New York.
[47] DI BARI, L.; LELLI, M.; PINTACUDA, G.; PESCITELLI, R.; MARCHETTI, F.; SALVADORI, P., *J. Am. Chem. Soc.* **2003**, *125,* 5549–5558.
[48] PERRIN, C. L.; GIPE, R. K., *J. Am. Chem. Soc.* **1984**, *106,* 4036–4038.
[49] MACURA, S.; ERNST, R. R., *Mol. Phys.* **1980**, *41,* 95.

Chapter 6
Yb for assigning 1,2-diol absolute configuration

In the previous chapters the NMR and NIR CD techniques were used to disclose the solution arrangement of several ytterbium complexes; this is possible thanks to the peculiar spectroscopic properties of the Yb(III) ion (Chapter 2 and 3). In this chapter we describe how these techniques can be applied to the determination of the absolute configuration of chiral 1,2-diols.

Chiral 1,2-diols are important building blocks in organic synthesis, because the hydroxylic function can be easily converted into several functionalities, which are present in many molecules of pharmacological interest. Furthermore, they can be synthesized in high enantiomeric excess (e.e.) from olefins through the Sharpless asymmetric dihydroxylation reaction [1]. Even if an empirical rule [1] allows one to predict the absolute configuration of the resulting diol from the structure of the precursor alkene and the nature of the chiral auxiliary, it lacks general validity.

Many methods have been developed for the determination of 1,2-diol absolute configuration (Section 6.1), and in particular those based on the CD measurements are rapid and sensitive. In contrast with those advantages, UV-Vis CD based methods may suffer from some limitations for substrates having strong absorption.

The alternative approach introduced in this chapter consists in observing the NIR CD spectra induced in the $f - f$ transitions of Yb when the diol is bound to the metal. A suitable choice of the Yb-complex allows one to obtain NIR CD spectra immediately after mixing the diol with the complex, with a good correlation between the sign of the Cotton effects and the absolute configuration of the diol. The methods proposed here were extended to a representative series of chiral diols, including *primary-secondary*, *secondary-secondary*, *primary-tertiary* diols, and also contemplating systems with several functionalities, sterical hindrances or strong UV absorption that can not be analyzed with other methods.

6.1. Determination of chiral 1,2-diols configuration

It is remarkable that the first molecule whose absolute configuration was determined is in fact a diol, namely (+)-tartaric acid. In the 1951 Bijvoet assigned it to (R,R) through anomalous X-ray scattering [2, 3], setting a milestone in the history of stereochemistry and confirming Fischer's and Rosanoff's postulate of the configuration of D-glyceraldehyde. A huge number of chemical correlations allowed the indirect configurational assignment of many other molecules.

This is a general non-empirical method, but it is limited to samples available as perfect crystals of suitable dimensions, and having at least one atom heavier than sulfur[1] [4]. For these reasons, more practical methods, based on spectroscopic measurements in solution (as circular dichroism), are widely used in the assignment of the absolute configuration. In the following, some of the most commonly used methods for the determination of the absolute configuration of diols based on CD are reported.

6.1.1. The diol di-benzoate method of Harada and Nakanishi

Probably the most used CD method for the assignment of the absolute configuration of cyclic diols is the dibenzoate method of Harada and Nakanishi [5]. It consists in the analysis of the CD observed in the benzoic 1,2-diester (or analogous derivatives) of the examined diol, which can be related in a non-empirical way to the diol absolute configuration.

The CD spectrum of the 1,2-benzoic diester shows a CD couplet due to the exciton coupling between the $\pi \to \pi^*$ electronic transitions of the benzoates, which is favored by the proximity of the two chromophores.

The amplitude, separation, and sign of the couplet (*i.e.* the sign of the highest wavelength band, see Figure 6.1) depends on the relative orientation of the components of the transition dipole moments that are directly related to the stereochemistry of the diol benzoates (Figure 6.1). the sign of the couplet is expected to be the same as the sign of the 1,2-diol chirality (*i.e.* the sign of the dihedral angle O−C−C−O) (Figure 6.1).

This method is particularly useful in the analysis of chiral cyclic diols, as in many common situations the conformation of the system can be predicted by inspection of simple molecular models, and the absolute configuration is easily deduced from CD [5]. For acyclic diols, the CD couplet can be related to the diol absolute configuration only after a de-

---

[1] The anomalous X-ray scattering is observed when heavy atoms are irradiated with X-ray close to the absorption frequency: tartaric acid was crystallized as sodium and rubidium salt. The aminoacids are generally crystallized as hydrobromides.

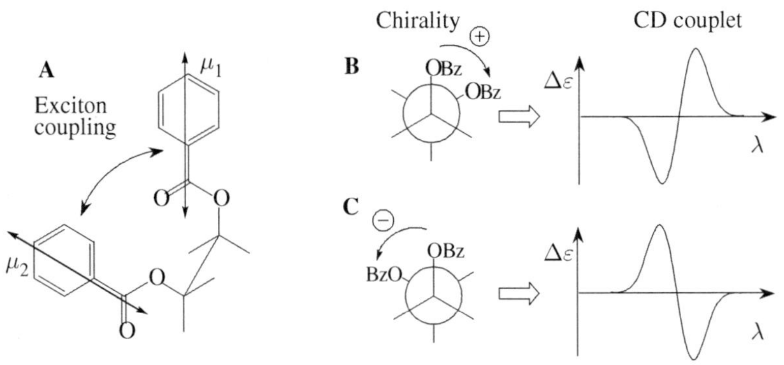

Figure 6.1. A) In the diol dibenzoate derivative, the electronic transition dipole moments μ_1 and μ_2 of the two chromophores are close enough to give exciton coupling. This interaction splits the electronic transition into two bands with opposite signs in the CD spectra (CD couplet). The sign of the couplet (*i.e.* the sign of the highest wavelength band) is directly correlated with the chirality of the diol conformer (**B, C**).

tailed conformational analysis, which often requires an investigation *ad hoc*. For this reason the method loses generality and the absolute configuration of acyclic diols is determined through various alternative methods, mainly based on the coordination of the diol to a metal complex.

6.1.2. Lanthanide diketonates method of Dillon and Nakanishi

Dillon and Nakanishi developed two practical methods based on the observation of CD of a solution of the diol with metal diketonate complexes [6]: in particular [Pr(dpm)$_3$] (dpm = dipivalomethanate) [7, 8], and [Ni(acac)$_2$] (acac = acethylacetonate) [9], (Figure 6.2).

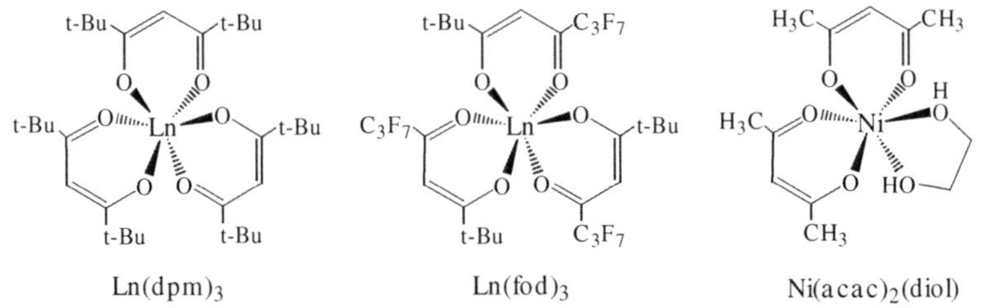

Figure 6.2. Structure scheme of the metal diketonate complexes used in the determination of the diol absolute configuration.

[Pr(dpm)$_3$] is a lanthanide complex, popular as NMR shift reagent; the three ligands symmetrically arrange around the metal with an octahedral coordination polyhedron, which may be capped by a water molecule [10].

The C_3 symmetry of the coordination polyhedron allows two specular Δ, Λ arrangements that interconvert rapidly in solution. When the chiral diol coordinates axially the metal it induces a preferential arrangement in the ligand cage. The Δ, Λ forms, which are equally distributed in the free complex, are unbalanced in the bound species with a prevalence of one form depending on the diol/complex interaction. Immediately after dissolving a chiral diol into a [Pr(dpm)$_3$] solution, the prevalent formation of a Δ or Λ form gives rise to an induced CD (ICD), which is characterized by an exciton couplet centred at 300 nm (Figure 6.3). This couplet is generated by the exciton interaction of the $\pi \rightarrow \pi^*$ transition of the diketonates, and has opposite sign for the two Δ or Λ ligand arrangements [9, 8].

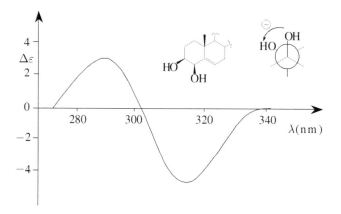

Figure 6.3. UV CD spectrum of a 1:1 mixture of cholest-5-ene-3β,4β-diol and [Pr(dpm)$_3$] 0.185 mM in dry CCl$_4$. The spectrum is recorded at room temperature 30 min after the mixing. The negative chirality of the diol moiety (reported in the picture) corresponds to the negative sign of the couplet.

Dillon and Nakanishi measured this ICD in presence of several cyclic and acyclic diols finding a general correlation between the sign of the couplet and the absolute configuration of the diol [7, 8]. On this empirical basis, a simple CD measurement can be used to assign the diol configuration without prior derivatization.

In the coordinated form, the diol is forced to assume *gauche* conformations (Figure 6.4), where its substituents occupy the axial or the equatorial positions with respect to the Pr$-$O$-$C$-$C$-$O ring. The sterical repulsion between these groups and the complex ligands forces the largest diol substituents to take the equatorial positions, and simultaneously induces a preferential Δ or Λ arrangement of the ligand.

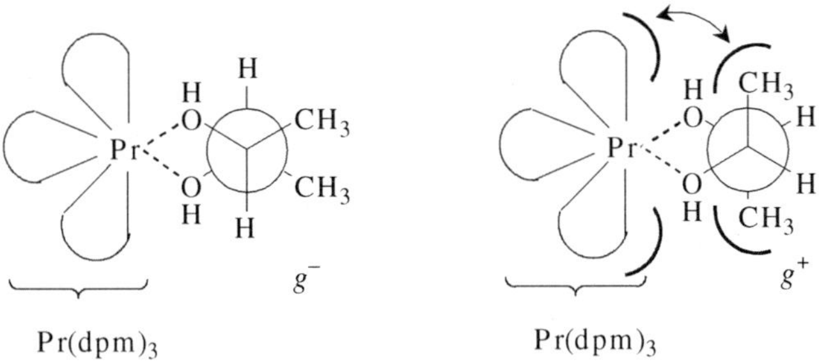

Pr(dpm)₃   Pr(dpm)₃

Figure 6.4. The *gauche* + (g^+) and *gauche* − (g^-) conformations of the coordinated diol give different interactions with the complex ligands. This favours the diol conformation with the largest substituents in the equatorial positions (g^- for the (R,R)-2,3-butane diol in the picture).

In this hypothesis diols that have similar structures coordinate the metal with the same conformation and induce the same ligand arrangement, with a consequent correspondence of the CD couplets. Unfortunately, the theoretical analysis can not be further extended, and this method remains based only on the empirical evidence.

The limited number of exception, the sensitivity and the ease of this method are its main advantages, but to give reproducible results, special care must be taken in drying both solvent (CCl_4 or $CHCl_3$) and diol. If chloroform is used, it must be freshly distilled and ethanol free: even traces of water promote the complex degradation with possible misleading CD spectra [8].

To circumvent this problem, Partridge [11, 12] first introduced [Eu(fod)₃] (fod = 6,6,7,7,8,8,8-heptafluoro-2,2-dimethyl-3,5-octanedioate), a fluorinated analogue of [Eu(dpm)₃] (Figure 6.2): the increased acidity of the fluorinated fod ligands makes the complex less sensitive to the water. Starting from the work of Partridge [11], [Eu(fod)₃] was used in the assignment of the absolute configuration of a variety of cholestan pentols [13, 14], and in the study of amino-alcohols [15], but it lacked a detailed study that extended the [Ln(dpm)₃] method to the fod complexes.

An analogous correlation between CD spectra and diol configuration is observed using the diketonate Ni(acac)₂ instead of Pr(dpm)₃ [9]. The main advantage of this method is the possibility of following the diol ICD both in the diketonate transition (300 nm) and in the $d-d$ electronic transition of the Ni(II) (around 600 nm), with the advantage of a double validation of the empirical correlation. On the other side, this method requires much larger diol quantities, and the CD correlation (again empirical) can not be extended to cyclic and sterically hindered substrates.

6.1.3. The di-molybdenum tetraacetate method of Snatzke

By far the most convenient, reliable and versatile empirical method for assigning the absolute configuration of 1,2-diols is the one proposed by Snatzke and Frelek and employing dimolybdenum tetraacetate $[Mo_2(AcO)_4]$ (DMTA) [16]. DMTA is a yellow crystalline compound, sufficiently stable at air and soluble in DMSO. The crystalline structure is endowed of D_{4h} symmetry [17], where a quadruple bond connects the two Mo atoms, and the four acetates chelate the Mo_2 unit (Figure 6.5A). By mixing a solution of DMTA with a chiral 1,2-diol an induced CD spectrum with several Cotton effects between 250 and 650 nm is obtained (Figure 6.5B), whose signs correlate with the absolute configuration of the substrate.

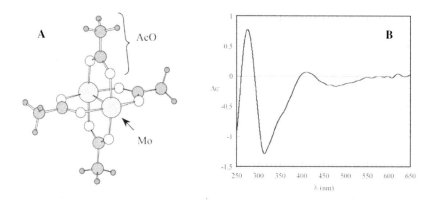

Figure 6.5. A) Structure of the $[Mo_2(AcO)_4]$ complex. **B)** UV CD spectrum of the mixture of $[Mo_2(AcO)_4]$ (1.43 mM) and (R,R)-2,3-butanediol (1.23 mM) in DMSO (T = 32 °C).

The ease of the procedure is one of the best qualities of this method: mixing very small amounts of diols (about 10 μ mol) in commercial (not necessarily dried) DMSO, at room temperature and open to air, leads in most cases to an immediate and intense CD spectrum. In principle, the diol can be used without derivatization, but DMTA does not inter-act selectively only with the 1,2-diol group: other systems having 1,1 or 1,2 or 1,3 donating atoms (as 1,3 diols, carboxylic acids, amino alco-hols, diamines etc.) may bind the metal center [16]: hence, special care must be used in presence of competing functionalities. The observed CD spectrum slowly decays several minutes after the mixing of the DMTA with the diol, this result is a consequence of the degradation of the active species.

Solution study simultaneously conducted with CD and NMR revealed the importance of traces of water (commonly present in the DMSO) in the formation of the active species, and provided a much more complete description of the equilibria occurring in the sample and of the structure of the CD-active species [18]; nevertheless at the moment the theoretical description of the origin of the various Cotton effects is still lacking.

Recently, a systematic study was extended over 19 chiral 1,2-diols especially designed to represent a wide spectrum of possible substrate [19]. Almost all the combinations of primary, secondary and tertiary diols were included, and no exception to the general correlation was found. Only in the case of the anthryl derivative **17** (Figure 6.6) a completely different CD spectrum was obtained, which can not be correlated with the other cases. This occurs because of the strong coupling between the electronic transitions of the complex with those of the substrate. This seems to constitute the only limitation to the application of this method.

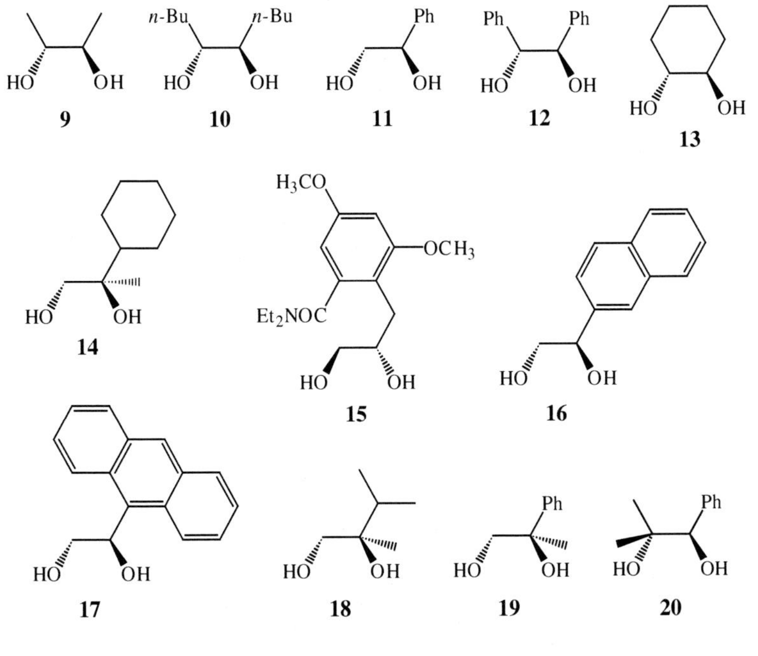

Figure 6.6. Structures of the diols used.

6.2. A novel approach to the lanthanide diketonate method

In analogy with [Pr(dpm)₃] above described (Section 6.1.2), we tested [Yb(dpm)₃] (**7**) as probe for the determination of the absolute configuration of diols. In presence of chiral 1,2-diols, solution of **7** shows induced UV CD spectra with couplets similar to those of [Pr(dpm)₃] [20]. Anyway, these systems are very sensitive to the presence of traces of water [8] that are responsible for a strong dependence of the CD on the sample preparation and measurement conditions [8, 20]. This is particularly inconvenient because it requires freshly dried solvent ($CHCl_3$, $CDCl_3$, CCl_4) and purified diols.

To circumvent this problem, we switched to [Yb(fod)₃] (Figure 6.2, see also Section 6.1.2), and we observed that the UV CD and NIR CD spectra of [Yb(fod)₃]/diol mixtures are indefinitely stable (even after weeks, and even in presence of small amounts of water), thus [Yb(fod)₃] appeared suitable for a detailed revision of the UV CD correlation (Section 6.2.1) and for an investigation of the nature of the active species (Section 6.2.3) [21]. Furthermore, taking advantage of the optical properties of the Yb(III) ion (Chapter 3), we first described the extension of the CD correlation to include NIR CD data as well, extending the configurational analysis to diols with strong UV absorptions (Section 6.2.2) [21].

6.2.1. Correlation between UV CD spectra and the absolute configuration of 1,2-diols using [Yb(fod)₃]

In Figure 6.7 the induced UV CD spectra of [Yb(fod)₃] (**8**) in presence of 1 equivalent of chiral diol (**9-14**, Figure 6.6) are reported.

The spectra show exciton couplets between 350 and 250 nm with two intense and opposite Cotton effects at 280 and 300 nm. The sign of the Cotton effect at 300 nm is univocally negative for a variety of substituted chiral diols, and correlate with the (R) or (R,R) configuration of the examined diols. This confirms that the method introduced with [Pr(dpm)₃] can be extended to the fod complexes.

With respect to the ICD of [Pr(dpm)₃] samples (Section 6.1.2) [8], the ICD spectra with **8** show strongly asymmetric couplets, where the Cotton effect at 300 nm is more intense than the one at 280 nm, and often flanked with a more or less pronounced band around 330 nm. It should be reminded that the asymmetry of the fod ligand, in contrast with the dpm, gives rise to different *fac* and *mer* isomers. These isomers have different orientation of the electric dipole transition moments, which in turn originate different exciton interaction, which may be the reason of the asymmetric profile of the resulting spectrum, but no further investigation was performed in this sense.

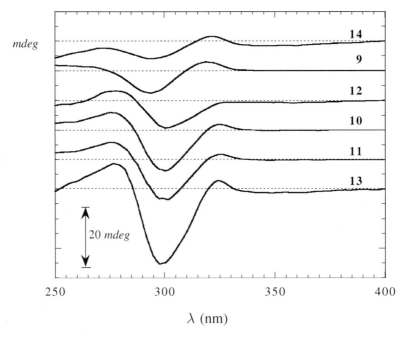

λ (nm)

Figure 6.7. UV CD spectra of diols **9-14**, with **8**: molar ratio 1:1, in anhydrous CCl$_4$ 0.35 mM. Path lenght 0.1 cm, room temperature. The e.e.'s of the used diols were >96%, only for **14** e.e. was 70% (Appendix A).

6.2.2. Correlation between NIR CD spectra and the absolute configuration of 1,2-diols using [Yb(fod)$_3$]

The coordination of the chiral diol induces discrimination between the Δ and Λ arrangement of the diketonate ligands around the metal (Section 6.1.2) [8]. This process can be followed through NIR CD spectroscopy, as the two opposite Δ and Λ coordination polyhedrons induce CD of opposite sign in the $f-f$ electronic transitions of ytterbium.

The NIR CD spectra acquired on several 1:1 [Yb(fod)$_3$]/diol mixtures show comparable spectra having a strong negative band at 976 nm, whose sign can be correlated with the (R) or (R,R) diols configuration (Figure 6.8). Interestingly, this correlation can be extended to molecules with different functional groups as **15**, or to diols with strong absorptions in the UV as **16** and **17**. In particular, the absolute configuration of the diol **17** can not be determined with any other traditional UV CD method (Section 6.1.1-3), included the one based on [Mo$_2$(AcO)$_4$] [19].

There are however two notable exceptions using the 2,3-butane diol **9** (Figure 6.10) and the 5,6-decane diol **10**, that give positive band at 978 nm. These two cases can be easily recognized from the other ones for

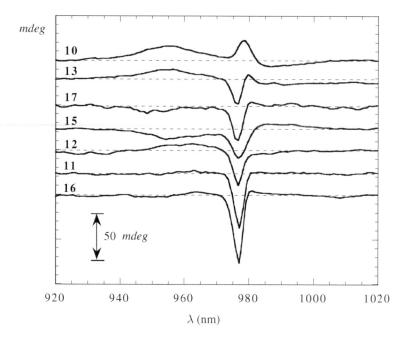

Figure 6.8. NIR CD spectra of diols **10-13**, **15-17** with **8** (5 mM, molar ratio 1:1, path length 10 cm). In order to make the comparison easier, the spectrum of diol **15** (having opposite configuration) was inverted, while those of molecules **10**, **13** and **17** have been multiplied by a factor 2. The e.e.'s of the used diols were >96%, only for **15** e.e. was 68% (Appendix A).

the different wavelength of the principal band (978 nm instead of 976 nm). Other weak bands may be observed around 960 nm but they are very noisy and uncorrelated with the diol configuration.

The different spectra recorded for the diols **9** and **10** may be a consequence of the small dimensions (**9**) or the extreme flexibility (**10**) that let these molecules arrange around the metal in a slightly different manner with respect to the other diols. Because NIR CD depends on the crystal field around the metal, it is much more sensitive than UV CD to changes in the ligand coordination.[2] The diols **9** and **10** may orient the diketonate helicity in the same manner as the other diols, inducing an UV CD couplet of the same sign, but they can also arrange differently in the coordination sphere of ytterbium, yielding consequently a different NIR CD spectrum.

---

[2] In Figure 6.10 it is apparent how the NIR CD is more sensitive than UV CD to the formation of a second diol coordination species [Yb(fod)$_3$(**9**)$_2$].

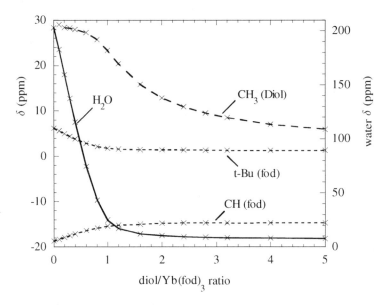

Figure 6.9. ^{1}H NMR shifts of selected protons as a function of diol/[Yb(fod)$_3$] molar ratio. The relative error on the shifts is essentially due to the linewidth and it is always below 2%.

6.2.3. The solution equilibra in the mixture of [Yb(fod)$_3$] with the chiral (2,3)-butane diol

The ^{1}H NMR spectrum of the [Yb(fod)$_3$] solution[3] shows only two peaks (t-Bu and CH protons), corresponding to an *effective C$_3$ symmetry* of the complex: this is not trivial considering the asymmetric nature of the fod ligand. A further broad peak at 203 ppm is assigned to the water traces that exchange between the free and the bound form in fast regime. A preliminary titration of [Yb(fod)$_3$] with water estimated the shift of the bound H$_2$O to around 250 ppm.

Upon addition of the dry chiral diol **9** (*e.g.* 2,3-butane diol), the aggregate [Yb(fod)$_3$(**9**)] is formed, and the diol resonances are strongly downfield shifted by the paramagnetic interaction (Figure 6.9). The diol is in fast exchange, on the NMR time scale, between the free and bound form so, in the course of the titration, its resonances move toward those of the free diol, following the increasing molar fraction of this species.

By coordinating to the complex, the diol replaces the bound water, whose resonance progressively drifts toward the chemical shift of the free

---

[3] This and the following NMR spectra are acquired at 298 K in CDCl$_3$, and at 7.1 T (300 MHz of ^{1}H Larmor frequency).

molecule. Also the shifts of the diketonate ligand change upon the diol coordination reflecting the structural changes in the new paramagnetic species [21].

A parallel titration was followed through NIR CD and UV CD spectroscopies (Figure 6.10): both techniques revealed the formation of a second bound species (2:1 diol/[Yb(fod)$_3$]), especially in excess of diol. The UV CD shows a small but noticeable change from 1 to 9 equivalents of added diol, at variance, the NIR CD spectra evidences better this process with the progressive formation of a negative peaks at 976 nm on going from 1 to 4 equivalent of added diol.

It is noteworthy that the addition of 1 equivalent of a monodentate achiral alcohol (1-butanol) to the 1:1 [Yb(fod)$_3$]/**9** mixture, produces the same variation in the NIR CD spectrum observed with the addition of the second equivalent of **9**.

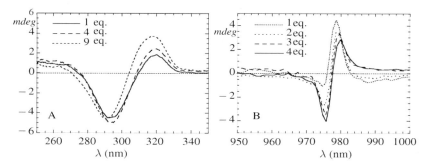

Figure 6.10. A) UV CD spectra of [Yb(fod)$_3$] after addition of different molar ratios of **9** (solvent: anhydrous CCl$_4$, path lenght 0.1 cm, 0.18 mM). **B)** NIR CD spectra of **8** after addition of different molar ratios of **9** (solvent: anhydrous CCl$_4$, path lenght 10 cm, 4.8 mM).

In order to interpret these results, we suppose that the second coordinated diol is bound as a monodentate ligand, with CN = 9 for Yb;[4] alternatively, it is also possible that in the [Yb(fod)$_3$(**9**)$_2$] complex both diols bind the metal like monodentate alcohols (with CN = 8) [8], or even that water participates to the metal coordination (equilibrium [6.4]). Thus, in the [Yb(fod)$_3$]/**9** mixture, several equilibria are simultaneously involved:

$$[Yb(fod)_3] + H_2O \rightleftharpoons [Yb(fod)_3(H_2O)] \qquad [6.1]$$

$$[Yb(fod)_3] + \mathbf{9} \rightleftharpoons [Yb(fod)_3(\mathbf{9})] \qquad [6.2]$$

---

[4] The hypothesis of two diol molecules both cheating the metal is not accepted because, up to now, no experimental evidences can justify CN = 10 for an ytterbium diketonate complex.

$$[Yb(fod)_3(\mathbf{9})] + \mathbf{9} \rightleftharpoons [Yb(fod)_3(\mathbf{9})_2] \qquad [6.3]$$

$$[Yb(fod)_3] + \mathbf{9} + H_2O \rightleftharpoons [Yb(fod)_3(\mathbf{9})(H_2O)] \qquad [6.4]$$

The equilibria [6.1] and [6.2] describe the formation of the water bound and diol bound species. The formation of the second species is described by [6.3], while [6.4] indicates the possible interaction of water with $[Yb(fod)_3(\mathbf{9})]$.

The quantitative analysis of the NMR titration between 0 and 1 eq. (where [6.1] and [6.2] dominate) allows us to calculate as $(7 \pm 1) \, 10^2 \, M^{-1}$ the affinity constant of $[Yb(fod)_3]$ for $\mathbf{9}$ [6.2].[5] From the analysis of the NMR spectra of $[Yb(fod)_3]$ with the diols $\mathbf{10}$, $\mathbf{11}$, $\mathbf{12}$, the affinity constants were roughly estimated of the same order of magnitude as for $\mathbf{9}$. The NIR CD analysis indicates that the dicoordinated complex $[Yb(fod)_3(diol)_2]$ is not formed with the more hindered diols $\mathbf{11}$ or $\mathbf{13}$. It is possible that the smaller dimensions of $\mathbf{9}$ let it have access to a further coordination site in the crowded sphere around ytterbium.

6.2.4. Solution structural analysis of the $[Yb(fod)_3(diol)]$ adduct

The quantitative analysis of the NMR titration of $[Yb(fod)_3]$ with $\mathbf{9}$ allows us to extrapolate the ligand and diol shifts in the bound form (Table 6.1). The bound complex $[Yb(fod)_3(\mathbf{9})]$ lacks any symmetry, so the three observed diol resonances (CH_3, CH and OH) evidence a fluxional structure in which the positions of ligand and diol rearrange dynamically ensuring an *effective* C_3 symmetry.

As $[Yb(fod)_3(\mathbf{9})]$ has no axial symmetry the calculation of the pseudo-contact shift should require the general equation (2.24), which contains also the rhombic term. However, the rapid rotation of the diol along the C_3 axis, as well as a fast equilibrium (on the NMR time scale) between the bound and free diol, *in fact* averages the positions of the diol around an *effective* C_3 axis. In such a case, it can be demonstrated that equation (2.24) is averaged to an *effective axial symmetry* and the rhombic term is ruled out [22], and (2.25) can be correctly used to the analysis of the pseucontact shift in $[Yb(fod)_3(\mathbf{9})]$. On this basis, we deduce that the diol (downfield shifted) occupies the axial position, and the ligands (upfield shifted) are equatorially distributed around the metal (Figure 6.11).

---

[5] We express the affinity constant as a concentration constant, which is equal to the thermodynamic stability constant under the assumption of unitary activity coefficients and using 1 M solutions as reference state.

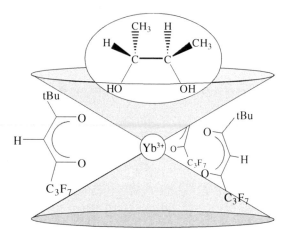

Figure 6.11. Schematic model of the (R,R)-2,3-butanediol (**9**) bound to a molecule of $[Yb(fod)_3]$. Protons inside the double cone are paramagnetic downfield shifted, outside these cones are upfield shifted.

The analysis of the NMR data can go further, determining which of the two possible g^+ and g^- diol conformations is involved in the chelation (Figure 6.4). The stereochemistry of the chelation is of crucial importance because it influences the Δ and Λ arrangement of the diketonates and consequently the induced CD (Section 6.1.2). This piece of structural information is deduced analyzing the pseudocontact shifts of the bound 2,3-butane diol previously obtained from the NMR titration (Section 6.2.3). Using MM2 calculations [23], two distinct g^+ and g^- models of the conformers of **9** bound to the metal were constructed, by setting the Yb–O distance to 2.5 Å [24].

From each model, a set of geometrical factors (GF) was derived and finally fitted against the experimental pseudocontact shift: the agreement of the g^- conformation with the experimental data is largely better than for the g^+ form (Table 6.1). Moreover, the good agreement factor ($R = 4.1\%$) supports the quality of the model.

Another validation of the g^- diol conformation comes from the CD data: the correlation among the NIR and UV CD spectra of several $(R)/(R,R)$ diols with the spectrum of the cyclic diol **13** (for which only the g^- conformation is possible in the (R,R) enantiomer) indicates that this conformation is generally preferred.

This is the first direct evidence of the diol chelating conformation, and strongly supports the hypothesis that the repulsions between the larger diol substituents and the ligand cage are dominant and determine the Λ/Δ arrangement of the ligand around the metal (see also Section 6.1.2).

Table 6.1. Experimental (δ_{exp}^{pc}) and calculated (δ_{calc}^{pc}) ^{1}H pseudocontact shifts for the diol **9** bound to [Yb(fod)$_3$]. The δ_{calc}^{pc}'s are calculated considering the two possible g^- and g^+ diol conformations; and GF are the corresponding geometrical factors (see Section 2.1.2). The $R(\delta^{pc})$ terms are calculated following equation (2.53).

Proton	δ_{exp}^{pc}	g^- Conformation			g^+ Conformation		
		$GF(\times 100)$ Å^{-3}	δ_{calc}^{pc}	$\delta_{exp}^{pc} - \delta_{calc}^{pc}$	$GF(\times 100)$ Å^{-3}	δ_{calc}^{pc}	$\delta_{exp}^{pc} - \delta_{calc}^{pc}$
CH	68.4	3.21	69.5	−1.2	2.24	64.4	4.0
CH$_3$	27.9	1.16	25.1	2.8	1.26	36.0	−8.2
OH	83.7	–	–	–	–	–	–
$\mathcal{D}$ (ppm Å^3)		2170	±	90	2870	±	350
$R(\delta^{pc})(\%)$		4.1			12.3		

6.3. New method based on the ytterbium triflate

The use of the ytterbium diketonate complex [Yb(fod)$_3$], described in Section 6.2, opened the way to the application of NIR CD to the determination of the absolute configuration of chiral glycols, especially for diols with strong UV absorptions that might not be analysed with the UV CD methods. In our alternative approach, the chromophore is the metal ion, and the observed NIR CD may arise from the intrinsic chirality of the coordination polyhedron, imposed by diol chelation. The same situation can take place for any complex where Yb^{3+} is coordinated to a chiral diol, without restrictions on the spectroscopic properties of the other ancillary ligands.

In order to explore this field, we considered the basic case where the complex derives from simple ytterbium salts as halides (YbX$_3$), triflate (Yb(OTf)$_3$) or oxide (Yb$_2$O$_3$), mixed with an excess of chiral diols. The diol establishes an equilibrium between bound and free species and NIR CD is sensitive only to the diol-containing species (Section 6.3.2). The basis of this method is to observe that the NIR CD spectrum is highly conserved when different diols (with the same configuration) are used, thus we can establish a correlation between the NIR CD spectrum and the diol absolute configuration.

With respect to more elaborated Yb complexes, inorganic salts are stable, accessible and inexpensive, and tolerate the presence of comparably large amounts of water or impurities as alcohols or amines. Excluding

Yb_2O_3, halide and triflate are quite soluble in many organic solvents, making a homogeneous phase with the diols (see Section 6.3.1). Furthermore, this method does not require substrate derivatization and is not destructive: in many cases the diol can be recovered washing away ytterbium with water and extracting the diol in the organic phase.

The nature of the optically active species was investigated through the combined use of NIR CD and multinuclear NMR (^{1}H, ^{19}F) (Section 6.3.3). The fallouts of this study are not limited to the method for the assignment of the diol chirality, but may be extended also to the application of the ytterbium ion in chemical synthesis [25]. Chiral diols may provide chiral auxiliaries that arrange around the metal and generate the environment needed for stereoselective reactions.

6.3.1. Choice of the solvent and of the ytterbium salt

The formation of the Yb/diol complex is favored using salt and solvent that enhance the Lewis acidity of the metal and do not interfere with the diol chelation (equilibrium [6.5]).

$$YbX_3 + \text{diol} \rightleftharpoons [Yb(diol)]X_3 \qquad [6.5]$$

Anhydrous ytterbium chloride ($YbCl_3$) or pseudo-halide ($Yb(OTf)_3$) dissolve well in dry methanol and gives weak but stable NIR CD spectra in the presence of (R,R)-2,3-butane diol (**9**). Methanol is a strong competitor for the chelating diol, and a significant excess of diol must be added to observe an intense CD spectrum. Less coordinating solvents are preferred, as they are less competitive towards the diol coordination. The best combination was found using acetonitrile (CH_3CN) or nitromethane (CH_3NO_2) as the solvent, and $Yb(OTf)_3$ as the ytterbium salt. The latter is much more soluble than ytterbium halides in non-protic solvents: so, using $Yb(OTf)_3$ stronger CD spectra are observed. The success of acetonitrile and nitromethane lies in their high dielectric constants ($\varepsilon/\varepsilon_0 = 36.0$ and 38.6 for CH_3CN and CH_3NO_2, respectively) [26] joined with a small or negligible attitude to coordinate lanthanides. Hence, they dissolve well both diol and $Yb(OTf)_3$ without solvating them, and favouring the reciprocal interaction.

6.3.2. Correlation between the 1,2-diols absolute configuration and NIR CD spectra

The shape of the NIR CD spectra of the mixture $Yb(OTf)_3$/**9** is conserved in all the investigated solvents (CH_3CN, CH_3NO_2, CH_3OH, THF, pyridine, $CHCl_3$, CH_2Cl_2), the difference in the intensity of the spectra also reflects different formation constants of the complexes (Figure 6.12).[6] Several diols are tested dissolving 4.0 diol equivalents in a 20 mM solution of $Yb(OTf)_3$ in CH_3CN; Figure 6.13 reports the corresponding spectra.

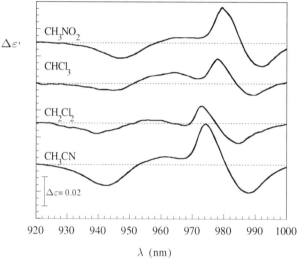

Figure 6.12. NIR CD spectra of $YbX_3/9$ mixture in different solvents. In CH_3CN and CH_3NO_2 $X^- = TfO^-$, while in $CHCl_3$ and CH_2Cl_2 $X^- = Cl^-$. In all the cases, the $YbX_3/9$ molar ratio was 1:4, the concentration of $[Yb^{3+}] = 20$ mM ($\Delta\varepsilon'$ is normalized to $[Yb(TfO)_3]$).

In spite of differences in the spectrum intensity, the appearance of position and sign of the CD bands are conserved for diols with the same absolute configuration. (R,R)-2,3-butandiol **9** gives the most intense spectrum, in which five different region can be distinguished (Figure 6.14).

The most intense signals are the band at 976 nm (III, positive for R diols), and the negative bands around 940 (I) and 990 nm (IV). Less intense are the positive Cotton effects at 960 (II) and 1010 nm (V). In these five regions fall most of the twelve transitions expected for Yb(III) ion (Section 3.1.2), which overlap to give the observed bands. The bands I, III, and IV conserve their sign for all the examined (R) or (R,R) diols, and they are the main diagnostic regions. On the contrary, the band II (950-970 nm) shows large intensity variations.

In particular, band II may be comparable of even larger than band III, as for **12**, **13**, and **20**, which are sterically hindered diols. We define this situation as A-type spectrum. Diols with a primary hydroxyl, as **14**,

---

[6] Small shifts of the band wavelengths in different solvent are observed, but they do not substantially change the shape of the spectrum.

18, **11** and **19**, or even 2,3-butane diol **9**, which are considerably more flexible, give rise to a NIR CD with band II definitely smaller than band III, which we call B-type.

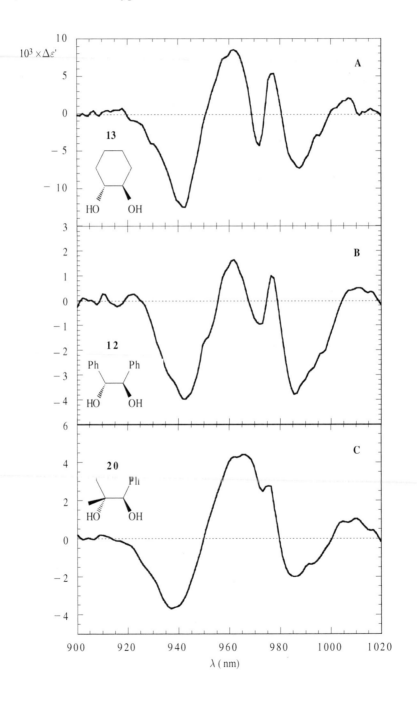

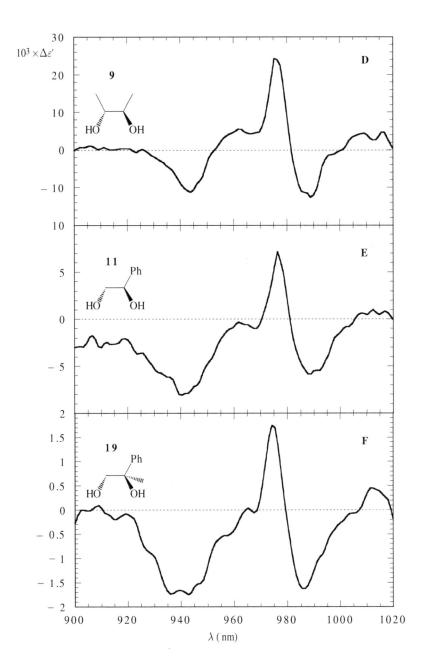

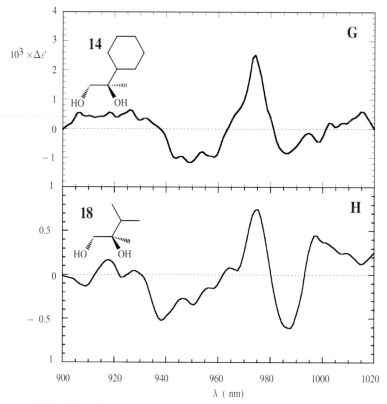

Figure 6.13. NIR CD spectra of the Yb/diol mixture with several diols: A) **13** (e.e. 99%), B) **12** (e.e. 99%), C) **20** (e.e. 89%), D) **9** (e.e. 99%), E) **11** (e.e. 99%), F) **19** (e.e. 56%), G) **14** (e.e. 70%), H) **18** (e.e. 30%). Each sample was prepared adding the diol in a 4:1 molar ratio with respect to Yb (0.5 ml, [Yb(TfO)$_3$] = 20 mM in commercial CH$_3$CN, path-length was 1 cm, $\Delta\varepsilon'$ is normalized to the Yb(TfO)$_3$ concentration). The number of scans was varied from 4 to 32 depending on the signal to noise. The spectra A-C have the characteristic "A-type" spectra where spectra D-H show a "B-type" profile (see in the text).

To test the applicability of this method to non enantiomerically pure diols, NIR CD spectra were recorded using (*R,R*)-2,3-butane diol (**9**) at variable enantiomeric excess (e.e.) from 0-100% (Figure 6.15). We observed no apparent variation of the spectrum profile with the band intensities that are directly proportional to the e.e. of the diol.

The linear dependence on the band intensity with the diol e.e., the presence of three coordinated diols (see Section 6.3.3), and the conserved spectrum profile indicate an high level of stereoselection in the diol binding with no appreciable formation of heterochiral species.

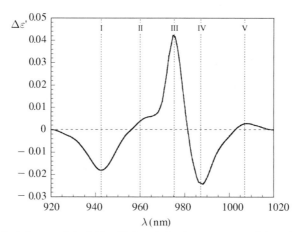

Figure 6.14. Scheme of the NIR CD bands in the spectrum of the Yb(OTf)$_3$/diol mixture (Yb/**9** ratio = 1:20, 60 mM Yb(OTf)$_3$ in CH$_3$CN). Roman numbers indicate the principal NIR CD bands.

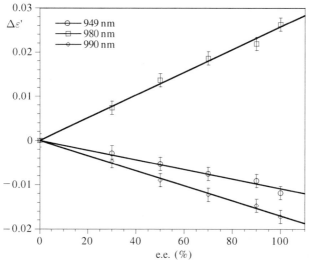

Figure 6.15. Plot of the NIR CD band intensity of the Yb(TfO)$_3$ and **9** at variable e.e. The intensity are measured at 949, 980, 990 nm with a Yb/diol ratio 1:4; [Yb(OTf)$_3$] = 5.25 mM in CH$_3$NO$_2$, path length 10 cm, T = 298K, $\Delta\varepsilon'$ is normalized to the Yb(TfO)$_3$ concentration.

6.3.3. Nature of the optically active species

The NIR CD spectra in the Yb(OTf)$_3$/diol mixture are the direct evidence of the formation of the Yb-diol complex. As no CD was observed upon addition of large excess of a chiral monoalcohol ((S)-2-butanol), one must deduce that only the rigid structure of the diol chelate is able to induce the asymmetric distortion in the whole coordination polyhedron.

To get insight into the nature of the active species, a titration of Yb(OTf)$_3$ with **9** was performed both in CH$_3$CN and CH$_3$NO$_2$ (Figure 6.16).

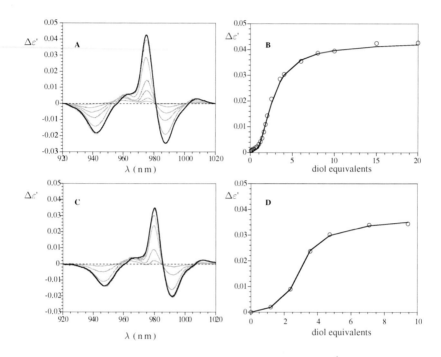

Figure 6.16. NIR CD titration of Yb(OTf)$_3$ with (R,R)-2,3-butane diol **9** in CH$_3$CN (A and B) and (CH$_3$NO$_2$) (C and D). The plots B and D report the trend of the intensity of the maximum around 980 nm as a function of the diol addition; dots indicates all the experimental points, and the continuous line the calculated points. For the sake of clarity, A shows only the spectra at 0.4, 1.0, 1.6, 2.0, 3.5, 10.0 eq. (gray line) and 20.0 eq. (dark line), and C shows only 1.18, 2.36, 3.54, 4.72, 7.08 eq. (gray line) and 9.44 eq. (dark line). Experimental conditions: A) [Yb(OTf)$_3$] = 60 mM, path length 1 cm, passing band 3.2 nm, time constant 0.5 s, room temperature; C) Yb(OTf)$_3$ 10.32 mM, path length 10 cm, passing band 3.2 nm, time constant 2 s, room temperature. $\Delta\varepsilon'$ is normalized to the Yb(TfO)$_3$ concentration.

Comparing the two titrations, it is evident that the diol binding constant in CH$_3$NO$_2$ is much higher than in CH$_3$CN: one can observe that if 7 equivalents of diol are sufficient to fully titrate the Yb(OTf)$_3$ solution in nitromethane, in CH$_3$CN the same amount of complex is formed only with more than 20 equivalents.

The higher stability of the Yb/diol adducts in nitromethane is explained with the lesser solvatation of ytterbium: the Yb(OTf)$_3$ remains suspended in the CH_3NO_2 solution, and it progressively dissolves while the diol is added. This is troublesome in the quantitative analysis of the titration, as the first part of the CH_3NO_2 titration is affected by the non complete dissolution of the $Yb(OTf)_3$.

The sigmoidal profiles of the plot of the band intensity at 974 nm as a function of the added diol equivalents indicate that more than one chiral species is formed during the titration (Figure 6.16B,D). This is also evident comparing the spectra recorded at the beginning of the titration (0.6 eq.) with the one of the end (20.0 eq.), which shows a clear change in the spectrum profile (Figure 6.17A,B).

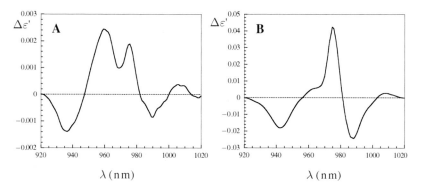

Figure 6.17. Profile of the NIR CD spectra at the beginning (**A**, 0.6 eq. of diol added) and the end of the titration (**B**, 20.0 eq.) of $Yb(OTf)_3$ with **9** in CH_3CN. See Figure 6.16A for experimental conditions.

It should be noticed that the NIR CD spectrum of the first species (Figure 6.17A) looks like the A-type spectra, typical for sterically hindered diols; this suggested us to do a full titration of $Yb(OTf)_3$ with the chiral cyclohexan diol **13** in order to observe the spectrum changes in the presence of a large excess of diol.

Indeed, by titrating $Yb(OTf)_3$ with **13** up to 17 diol equivalents (in CH_3CN), the spectrum progressively evolves passing from a A-type spectrum (4-6 eq.) to a B-type spectrum in large diol excess (Figure 6.18).

A detailed analysis of the initial points of the titration shows that another species is formed around 3 eq. of added diol. Overall three diol-bound species are formed along the titration as it is apparent by plotting the intensity of the principal Cotton effects as a function of the equivalents of **13**: the first dominates below 2 eq., the second between 2 and 5 eq. (A-type), and the latter (B-type) over 5 eq. of diol added (Figure 6.18B). The absence of an isosbestic point in Figure 6.18A as well as the factor analysis (data not shown), confirm the formation of three different species.

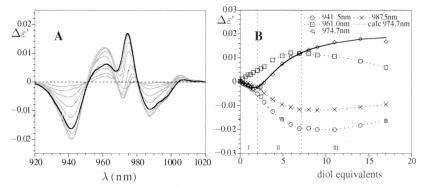

Figure 6.18. NIR CD titration of Yb(OTf)$_3$ with **13** in CH$_3$CN (A and B). For the sake of clarity, A shows only the spectra at 0.6, 1.2, 2.0, 2.8, 3.8, 5.9, 8.0, 11.0 eq. (gray line) and 17.0 eq. (dark line). In the plot B the intensities at 941.5, 961.0, 974.7, and 987.5 nm are reported. The continuous line joins the fitted points, and dotted lines join the experimental points. Three important slope changes (I, II, III) are reported. Experimental conditions: [Yb(OTf)$_3$] = 60 mM, path length 1 cm, band passing 3.2 nm, time constant 1.0 s, 8 acquisitions, room temperature. $\Delta\varepsilon'$ is normalized to the Yb(TfO)$_3$ concentration.

The resulting picture is a system of equilibria that point to the formation of three species: [Yb(diol)], [Yb(diol)$_2$], [Yb(diol)$_3$], where [Yb(diol)$_3$] corresponds to the B-type species, and [Yb(diol)$_2$] to the A-type ([6.6-8]).

$$Yb(OTf)_3 + diol \rightleftharpoons [Yb(diol)] \tag{6.6}$$

$$[Yb(diol)] + diol \rightleftharpoons [Yb(diol)_2] \tag{6.7}$$

$$[Yb(diol)_2] + diol \rightleftharpoons [Yb(diol)_3] \tag{6.8}$$

In this model, ytterbium is surrounded at maximum by three diol molecules, providing a CN = 6; this may look unusual for metals able to give CN up to 8 or 9 (see Chapter 1). ^{19}F NMR experiments on Yb(OTf)$_3$/**9** mixture, and in large excess of diol showed that the triflate ion is still partially bound to the metal (data not shown); it can not be excluded that also residual water participates to the metal coordination. Solvent coordination should be excluded for the [Yb(diol)$_3$] species because the NIR CD spectrum is exactly the same in different solvents, including non-coordinating media as nitromethane or chloroform (Figure 6.12). Thus, a more correct formula for the B-type species is [Yb(diol)$_3$(OTf)$_n$]$^{3-n}$, where n is the number of bound triflate ions, and 7 < CN < 9.[7]

---

[7] In principle, it can not be excluded that the diol-bound species are four, and that the first one (*e.g.* [Yb(diol)]) does not have an appreciable CD. Anyway, the absence of clear evidences in this sense, and the fact that triflate is coordinated reasonably excludes this possibility.

The formation constants K_1, K_2, K_3 for the equilibria [6.6-8], respectively, were fitted from the curves of the CD intensity in the butane diol and cyclohexane diol titrations (Figure 6.16, Figure 6.18, and Table 6.2).

Table 6.2. Formation constants[8] and molar CD's obtained through non-linear fit of the titration curves in Figure 6.16 and Figure 6.18. The molar CD's $\Delta\varepsilon_1$, $\Delta\varepsilon_2$, $\Delta\varepsilon_3$, and the formation constants K_1, K_2, K_3 are relative to the species [Yb(diol)], [Yb(diol)$_2$], [Yb(diol)$_3$], respectively. The molar CD's are measured at 974 nm in CH_3CN and 979 nm in CH_3NO_2. The agreement factor R is a Willcott-type term (Section 2.4).[9] The fitted constants K_1, K_2, $\Delta\varepsilon_1$, $\Delta\varepsilon_2$, measured in CH_3NO_2 suffer from the partial dissolution of Yb(OTf)$_3$ and are not reported.

Diol	Yb(OTf)$_3$/**9**		Yb(OTf)$_3$/**13**
Solvent	CH_3CN	CH_3NO_2	CH_3CN
K_1 (M^{-1})	$\sim 10^4$	—	$(1.4 \pm 0.4)10^2$
K_2 (M^{-1})	$(1.3 \pm 0.4)10^2$	—	$(3.3 \pm 0.9)10^1$
K_3 (M^{-1})	$(1.5 \pm 0.1)10^1$	$(1.9 \pm 0.1)10^2$	(7.1 ± 0.2)
$\Delta\varepsilon_1$ (M^{-1}cm^{-1})	$\sim 10^{-3}$	—	$(-2.7 \pm 0.5)10^{-3}$
$\Delta\varepsilon_2$ (M^{-1}cm^{-1})	$(1.4 \pm 0.1)10^{-2}$	—	$(-7.9 \pm 0.5)10^{-3}$
$\Delta\varepsilon_3$ (M^{-1}cm^{-1})	$(4.4 \pm 0.1)10^{-2}$	$(3.78 \pm 0.05)10^{-2}$	$(2.34 \pm 0.05)10^{-2}$
$R(\%)$	2.8	1.3	3.1

As above indicated, it is apparent that the binding constants in nitromethane are roughly one order of magnitude higher than in acetonitrile, with comparable molar CD's. In acetonitrile the sterically hindered diol **13** shows binding constants 2-4 times smaller than **9**. This is the reason why in the spectra of Figure 6.13, recorded at 4 equivalents of added diol, **13** shows an A-type spectrum due to the prevalence of the [Yb(diol)$_2$] species, while adding the same amount of diol **9** the [Yb(diol)$_3$], the B-type spectrum is prevalent.

The analysis of the titration curves of the diols **9** and **13** in CH_3CN allowed us to reconstruct the NIR CD spectrum of the species [Yb(diol)$_2$] and [Yb(diol)$_3$].[10] The estimated spectra, reported in Figure 6.19, show

---

[8] We report here the concentration constants of formation, which are equal to the stability constant under the assumption of activity coefficients equal to 1 (see Experimental Part in Appendix A) and taking as reference state solution with concentration 1 M.

[9] The agreement factor is defined as $R = \sqrt{\dfrac{\sum_i (Y_{exp;\,i} - Y_{calc;\,i})^2}{\sum_i (Y_{exp;\,i})^2}}$, where Y_{exp} and Y_{calc} are the experimental and calculated band intensities.

[10] The reconstructed CD spectrum of the [Yb(diol)] species is too weak and has too large errors compared to the species [Yb(diol)$_2$] and [Yb(diol)$_3$] thus, we do not report it.

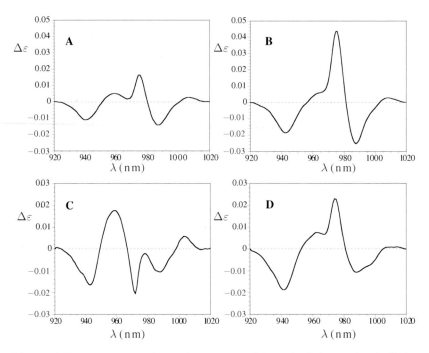

Figure 6.19. Reconstructed NIR CD spectra of the [Yb(diol)$_2$] and [Yb(diol)$_3$] species on the basis of the fitted binding constant (Table 6.2). A and B report the [Yb(diol)$_2$] and [Yb(diol)$_3$] spectra for the diol **9**, respectively. C and D, the [Yb(diol)$_2$] and [Yb(diol)$_3$] spectra for the diol **13**, respectively.

that [Yb(diol)$_3$] complexes generally have spectra more intense than [Yb(diol)$_2$]. It is apparent that the [Yb(diol)$_3$] species have essentially the same spectrum for both the diols **9** and **13**, with the B-type spectrum. The reconstructed spectra of the [Yb(diol)$_2$] species confirms that it has bands more or less at the same wavelength as those of [Yb(diol)$_3$], but with two notable differences: the band at 960 increases its relative intensity (for example compared to the band at 975 nm), and the sign of the band at 975 nm changes sign comparing [Yb(**9**)$_2$] and [Yb(**13**)$_2$].[11]

The strict correspondence of the NIR CD spectra of [Yb(**9**)$_3$] and [Yb(**13**)$_3$] indicates that ytterbium is surrounded by a similar coordination environment, and consequently we deduce that both complexes have the same CN and the diol arranges in the same conformation. Since the ring imposes the g^- conformation to **13**, also **9** should be arranged in g^-.

---

[11] These two aspects can also be appreciated looking the trend of the titrations at low diol/Yb molar ratio reported in Figure 6.16 and Figure 6.18. In particular is notable the different behaviour at low diol/Yb molar ratio between diols **9** and **13**.

In such a way, the (R,R)-diol minimizes the steric interaction of the diol substituents and the complex ligands (see also Section 6.1.2).

At variance, the spectra of the $[Yb(\mathbf{9})_2]$ and $[Yb(\mathbf{13})_2]$ evidences differences, and it is possible that they have different structures. In strict analogy with what observed for the NIR CD spectra of $[Yb((R)\text{-DOTAMPh})]$ and $[Yb((S)\text{-DOTAMNp})]$ in several solvents [27, 28] (see Section 3.2), the inversion of the band at 975 nm can be ascribed to a different influences of the solvent. In particular, it is possible that, differently from $[Yb(\text{diol})_3]$, the $[Yb(\text{diol})_2]$ species are solvent coordinated, and the different sterical hindrance of the diols $\mathbf{9}$ and $\mathbf{13}$ influences the number of bound solvent molecules, with a consequent effect on the NIR CD spectra. Anyway, no further evidences that confirm this hypothesis are available at the moment.

The ^{1}H NMR study of the $Yb(OTf)_3/\mathbf{9}$ titration is hampered by fast exchange (on the NMR time scale), between the free diol and the several diol-bound species. Even working a low temperature $(-40\ ^\circ\text{C})$, no signal-decoalescence was observed. During the titration, the diol chemical shift results from the weighted average of the shifts of the free and the bound forms: in practice the resonances progressively move towards those of the free diol, which is the prevalent form, in large diol excess (Figure 6.20).

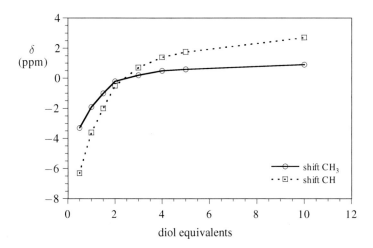

Figure 6.20. ^{1}H NMR shifts of CH_3 and CH protons of $\mathbf{9}$ in the titration of $Yb(OTf)_3$ in CD_3CN ($[Yb(OTf)_3] = 20$ mM, $T = 298$ K).

The complicated system of equilibria hampers the extrapolation of the shifts of the bound forms from the NMR titration. This prevents to obtain structural information from the NMR data (*i.e.* from pseudocontact shifts and relaxation times), and even about the symmetry of the formed species.

6.3.4. Summary of the method

Ytterbium triflate is a useful probe for the determination of the absolute configuration of diols through NIR CD spectroscopy. The method is practical and easy: the diol and $Yb(OTf)_3$ must be dissolved in CH_3CN (or CH_3NO_2) in molar ratio 4:1. If nitromethane is used (recommended for bulky diols), the salt dissolution can be accelerated using ultrasounds. Depending on the instrument sensitivity, it is generally suggested to use $Yb(OTf)_3$ solution 10 mM or more concentrated, to reduce the acquisition time and improve the signal-to-noise. NIR CD is acquired on the clear solution between 900 and 1020 nm: the bands at 940, 970, 990 nm (I, III, IV in Figure 6.14) are diagnostic: the sign alternation $-, +, -$, of these bands indicate an (R) or (R,R) absolute configuration of the diol. If observed, the other bands at 960 and 1010 nm can be used to further validate the assignment: taking as reference the spectra in Figure 6.13 and Figure 6.14, and the scheme summarised in Table 6.3. This method can also be applied to diols with variable enantiomeric composition (e.e. $< 100\%$), with no variation in the spectral pattern.

The spectrum profile may change according to the relative abundance of the $[Yb(diol)]$, $[Yb(diol)_2]$ and $[Yb(diol)_3]$ species.

Table 6.3. Summary of the method of the $Yb(OTf)_3$. For each Cotton effect the sign for the $(R)/(R,R)$ diol configuration for secondary and primary chiral diols are reported. The diagnostic bands are grey highlighted.

Band	λ (nm)	Sign (R,R)-sec-sec/sec-ter	Sign (R)-prim-sec/prim-ter
I	940	$-$ (intense)	$-$ (intense)
II	960	$+$ (intense)	$+$ (weak)
III	976	$+$ (intense)	$+$ (intense)
IV	990	$-$ (intense)	$-$ (intense)
V	1010	$+$ (weak)	$+$ (weak)
NIR CD spectra type for $Yb(OTf)_3$/diol ratio 1:4		A	B

In case of deviation from the typical spectrum (Figure 6.14) it is suggested to increase the diol molar ratio in order to have a larger prevalence of the diagnostic [Yb(diol)$_2$] and [Yb(diol)$_3$] species. The method is not destructive and the diol can be generally recovered by washing the sample with H$_2$O and by extracting the diol with an organic solvent.

6.4. Conclusions

In this chapter, Yb(III) NIR CD spectroscopy is used to determine the absolute configuration of chiral 1,2-diols. Ytterbium(III) acts as a chromophore, which is sensitive to the arrangement of the ligand around it. While in the previous chapters we demonstrated how the study of NIR CD spectra provides precious information about the chiral structure of the complex, here we use this technique to have indirect information about the absolute configuration of the ligand: in this case, a 1,2-diol. The correlation between the chirality of the ligand and the consequent chiral arrangement of the whole Yb(III)-diol adduct is the key-point, which allows us to assign the diol absolute configuration on the basis of the NIR CD spectrum of the adduct.

Two distinct methods for the determination of the absolute configuration of 1,2-diols are developed; both refer to the use of ytterbium as [Yb(fod)$_3$] or Yb(OTf)$_3$. These derivatives are able to interact with the diol and to generate NIR CD spectra. The shape and the sign of the bands are strictly related to the absolute configuration of the 1,2-diols used.

The [Yb(fod)$_3$] complex is proved to be a valid probe for the determination of the absolute configuration of chiral vicinal diols through CD measurement [21]. This complex represents an improvement of the [Pr(dpm)$_3$] method introduced by Nakanishi and Dillon [7, 8], offering two important advantages: *a*) [Yb(fod)$_3$] is much less sensitive to the presence of water traces, *b*) it allows one to use NIR CD measurement, as well as UV CD, to establish the absolute configuration of the diol. Indeed, the coordination of the chiral diol induces a preferential helicity in the arrangement of the diketonates responsible for an induced CD in both the UV and NIR regions. UV CD provides a very sensitive tool and no exception was found in the examined cases, although it may be limited to the cases where the diol does not contain absorbing chromophores. NIR CD is less sensitive, but it allows one to investigate diols as **16** and **17** that have strong absorptions in the UV region, and are less suited to other UV CD-based methods.

In many cases both UV and NIR CD measurements can be carried out on the same sample, providing independent validations of the absolute configuration of the diol, and minimizing the possible errors due to diols (see the case of **9** and **10**) that do not follow the general NIR correlation. Indeed, NIR CD is more sensitive than UV CD to variations in the coordination sphere of ytterbium, and may be that exceptionally small diols (as **9**) or flexible ones (as **10**), have a different coordination arrangement and give different NIR CD spectra [21].

The analysis of the structure of these systems is enriched by the NMR study: taking advantage of the paramagnetic properties of the Yb(III) ion, the arrangement of the chelating diol is determined, and definitively assigned to the g^- conformation (for $(R)/(R,R)$ diols), this confirms the hypothesis put forward by Nakanishi and Dillon and derived only from UV CD measurements.

The second NIR CD method is related to the use of the $Yb(OTf)_3$ salt. This salt dissolves in CH_3CN and CH_3NO_2, and in the presence of an excess of 1,2-diols form stable 2:1 and 3:1 diol/Yb complexes with a characteristic NIR CD spectra, that correlate well with the absolute configuration of the 1,2-diols. Even if the method is less sensitive than the previous one, no exceptions are found in the examined cases. Furthermore, the formed adducts are stable and are not sensitive to the presence of traces of water: so they can be obtained using commercial solvent without purification. The method is not destructive and generally the used diol can be easily recovered after the measurement. The nature of the active species was examined in detail through NIR CD titrations that allowed us to discern the formation of the 1:1, 2:1, 3:1 species, with the last one endowed with the most intense CD bands. Analogously to the case of $[Yb(fod)_3]$ the diol chelates the metal in the g^- conformation (for $(R)/(R,R)$ diols).

Both these studies represent complementary approaches that extend, but not exclude, the already known UV CD methods for the determination of the absolute configuration of 1,2-diols. The present methods have the advantage to be applied to diols with strong UV chromophoric groups, which can not be analyses with the UV CD methods in view of their strong absorption. Otherwise, they can be used together with the UV CD methods to have an additional and independent validation of the proposed absolute configuration.

References

[1] KOLB, H. C.; VANNIEUWENHZE, M. S.; SHARPLESS, K. B., *Chem. Rev.* **1994**, *94*, 2483.

[2] BIJVOET, J. M.; PEERDEMAN, A. F.; VAN BOMMEL, A. J., *Nature* **1951**, *168*, 271.

[3] BIJVOET, J. M., *Endeavor* **1955**, *14*, 71.

[4] ELIEL, L. E., WILEN, S. H., *Stereochemistry of Organic Compounds*, **1994**, Wiley-Interscience: NewYork.

[5] (a) HARADA, N.; NAKANISHI, K., *Circular Dichroic Spectroscopy – Exciton Coupling in Organic Stereochemistry*, **1983**, University Science Book:

Mill Valley. (b) NAKANISHI, K.; BEROVA, N., *Circular Dichroism: Principles and Applications*; NAKANISHI, K.; BEROVA, N.; WOODY, R. W. (eds.), **2000**, 2nd ed.; Wiley-WCH: New York, Chapter 12, 337–382.

[6] DILLON, J.; NAKANISHI, K., *J. Am. Chem. Soc.* **1974**, *96*, 4057–4059.

[7] NAKANISHI, K. DILLON, J., *J. Am. Chem. Soc.* **1971**, *93*, 4058–4060.

[8] DILLON, J.; NAKANISHI, K., *J. Am. Chem. Soc.* **1975**, *97*, 5417–5422.

[9] DILLON, J.; NAKANISHI, K., *J. Am. Chem. Soc.* **1975**, *97*, 5409–5417.

[10] COTTON, S., *Lanthanides & Actinides*, **1991**, Oxford University Press: New York.

[11] PARTRIDGE, J. J.; TOOME, V.; USKOKOVIC M. R., *J. Am. Chem. Soc.* **1976**, *98*, 3740–3741.

[12] PARTRIDGE, J. J.; SHIUEY, S.; CHADHA, N. K.; BAGGIOLINI, E. G.; BLOUNT, J. F.; USKOKOVI, M. R., *J. Am. Chem. Soc.* **1981**, *103*, 1253–1255.

[13] (a) DAYAL, B.; SALEN, G.; PADIA, J.; SHEFER, S.; TINT, G. S.; WILLIAMS, T. H.; TOOM, V.; SASSO G., *Chemistry and Physics of Lipids* **1992**, *61*, 271–281; (b) DAYAL, B.; KESHAVA R.; SALEN, G.; SEONG, W. M.; PRAMANIK, B. N.; HUANG, C. E.; TOOME, V., *Pure & Appl. Chem.* **1994**, *66*, 2037–2040.

[14] (a) DAYAL, B.; TINT, G. S.; SHEFER, S.; SALEN, G., *Steroids* **1979**, *33*, 327–338; (b) DAYAL, B.; SALEN, G.; TINT, G. S.; TOOME, V.; SHEFER, S.; MOSBACH, E. H., *J. Lipids Research.* **1978**, *19*, 187–190; (c) Dayal, B.; Salen, G.; Toome, V.; Tint, G. S. *J. Lipids Research.* **1986**, *27*, 1328–1332.

[15] (a) TSUKUBE, H.; HOSOKUBO, M.; WADA, M.; SHINODA, S.; TAMIAKI, H., *J. Chem. Soc. Dalton Trans.* **1999**, 11–12; (b) TSUKUBE, H.; SHINODA, S., *Enantiomer* **2000**, *5*, 13–22.

[16] (a) SNATZKE, G.; WAGNER, U.; WOLFF, H. P., *Tetrahedron* **1981**, *37*, 349–361. (b) FRELEK, J.; GEIGER, M.; VOELTER, W., *Curr. Org. Chem.* **1999**, *3*, 117–146 and references therein.

[17] COTTON, F. A.; MESTER, Z. C.; WEBB, T. R., *Acta Cryst.* **1974**, *B30*, 2768–2770.

[18] DI BARI, L.; PESCITELLI, G.; SALVADORI, P., *Chem. Eur. J.* **2004**, 1205–1214.

[19] DI BARI, L.; PESCITELLI, G.; PRATELLI, C.; PINI, D.; SALVADORI, P., *J. Org. Chem.* **2001**, *66*, 4819–4825.

[20] LELLI, M., Degree Thesis, **1999**, University of Pisa, Italy.

[21] DI BARI, L.; LELLI, M.; PINTACUDA, G.; SALVADORI, P., *Chirality* **2002**, *14*, 265–273.

[22] PETERS, J. A.; HUSKENS, J.; RABER, D. J., *Prog. Nucl. Magn. Reson. Spec.* **1996**, *28*, 283–350.

[23] CS CHEM 3D PRO v.3.5.2., Cambridge Soft Corporation; Cambridge, Massachusetts, 02139 USA.

[24] PARKER, D.; DICKINS, R. S.; PUSCHMANN, H.; CLOSSLAND, C.; HOWARD, J. A. K., *Chem. Rev.* **2002**, *102*, 1977–2010, and references therein.

[25] KOBAYASHI, S., *Lanthanides: Chemistry and Use in Organic Synthesis*, *Topics in Organometallic Chemistry*, **1999**, Springer-Verlag: Berlin.

[26] HUHEEY, J. E.; KEITER, E. A.; KEITER, R. L., *Inorganic Chemistry: Principles of Structure and Reactivity*, **1993**, 4th ed.; HarperCollins College Publisher.

[27] DI BARI, L.; PINTACUDA, G.; SALVADORI, P.; DICKINS, R. S.; PARKER, D., *J. Am. Chem. Soc.* **2000**, *122*, 9257–9264.

[28] DICKINS, R. S.; PARKER, D.; BRUCE, J. I.; TOZER, D. J., *Dalton Trans.* **2003**, 1264–1271.

Conclusions

The work of this thesis describes the structural study performed in solution on several ytterbium complexes, with a special regard for systems having interest in asymmetric catalysis. This work was carried out by a synergic use of NIR CD and NMR spectroscopies, taking advantage of the optical properties of the Yb(III) ion and of the paramagnetic contributions to NMR shift and relaxation rates.

In Chapter 4, these techniques were used to investigate two catalysts $(Na_3[Yb((S)\text{-BINOL})_3]$, $K_3[Yb((S)\text{-BINOL})_3])$ which belong to the larger family of lanthanide heterobimetallic catalysts. The analysis of the NMR data allowed us to obtain a refined solution structure of these systems that, in particular for $Na_3[Yb((S)\text{-BINOL})_3]$, differs from the one determined through X-ray diffraction. This demonstrates that the structural study in solution is an important complement of the information coming from the crystallographic structure. NMR and NIR CD have been effective also in the study of solution dynamics, providing a detailed picture of the ligand lability and evidencing the weak tendency of such systems to coordinate small molecules such water. This sheds new light on the behaviour of such systems and suggests a possible alternative hypothesis about the catalytic mechanism of these Yb-complexes.

The structural analysis performed on $[Yb((S)\text{-THP})]$, $[Yb(THED)]$, two Yb-DOTA-like complexes (Chapter 5), as well as the investigation of the $[Yb(fod)_3((R,R)\text{-2,3-butandiol})]$ adduct (Chapter 6), demonstrates that these techniques have a general applicability and can be extended to different systems. Furthermore, NIR CD and paramagnetic NMR are not only effective in the determination of the solution structure, but can be also used to analyse conformational and proton exchange equilibria, possibly involved in complex processes, like the dimerization of $[Yb((S)\text{-THP})]$. Their potential in investigating the structure and the stereochemistry of chiral Yb-systems is also manifest in the work described in Chapter 6, which culminates in two new and practical methods for the assignment of the absolute configuration of chiral 1,2-diols through NIR

CD measurement. These methods complement the already reported UV CD protocols, in so far as they allow one to analyse also molecules with chomophores with strong UV-Vis absorption and that can not be investigated otherwise.

Several considerations can be derived from the results of this thesis. First, the analysis of the NMR paramagnetic contributions, and in particular of the pseudocontact shifts, provides most of the quantitative information that enabled us to establish the solution structure of six molecules ($Na_3[Yb((S)-BINOL)_3]$, $K_3[Yb((S)-BINOL)_3]$, $[Yb((S)-THP)]$, $[Yb((S)-THP)]_2$, $[Yb(THED)]$, $[Yb(fod)_3((R,R)-2,3-butandiol)]$). Indeed, the pseudocontact interaction is unique, in so far as it directly relates NMR shifts to the position of the nucleus with respect to the metal. Even if analogous analyses could be carried out for several lanthanides, in Yb(III) complexes it is much easies to isolate structural restraints from the NMR data. The sensitivity of pseudocontact shifts to small structural changes, allowed us to monitor for example the rearrangement upon dissolution of $Na_3[Yb((S)-BINOL)_3]$ (Chapter 4), as well as the diol conformation in $[Yb(fod)_3((R,R)-2,3-butandiol)]$, (Chapter 6). The study of the $K_3[Yb((S)-BINOL)_3]/(R)$- and (S)-BINOL mixture or of the $[Yb(THED)]$ complex shows that the shift differences among conformers or exchanging species are so large that the exchange process appears slow on the paramagnetic NMR time scale. In such conditions, it was possible to investigate in detail dynamic processes, distinguishing the species involved and determining the exchanging rates.

A second consideration is that the pseudocontact coupling is a long-range interaction. We emphasise that the intermolecular pseudocontact shifts in the dimer $[Yb((S)-THP)]_2$ are responsible for consistent changes in the ^1H NMR spectrum with respect to the monomer; and it makes possible to describe the aggregation process and the rearrangement of the approaching molecules.

The third, but not less important aspect is the role played by NIR CD spectroscopy. Many of the structural results reported here would not be possible without the information coming from NIR CD. For example, this is the case of the assignment of the ligand conformation of $[Yb((S)-THP)]$ and the determination of the pK of proton dissociation (Chapter 5), as well as the disclosure of the $K_3[Yb((S)-BINOL)_3]/(R)$-BINOL exchange (Chapter 4), or the analysis of the $Yb(OTf)_3$/diol equilibria (Chapter 6). Two are the main advantages of this technique: the first is that NIR CD is sensitive and selective to chiral Yb-species, the second is that NIR CD reflects the chiral environment around the metal, and allows one to monitor the structural changes on the coordination sphere of the metal.

The symbiosis between NIR CD and NMR is here demonstrated: from one side, NMR offers quantitative restraints for structural optimisation, from the other side NIR CD offers a point of view on the chirality of the system, which is not appreciated by NMR.

NIR CD, which is sensitive to the chiral environment around Yb(III), can be accompanied by UV CD, which observes the optical transition of the ligand, and which is sensitive to the ligand chirality. The study of interaction among [Yb(fod)₃] and chiral diols reported in Chapter 6, well demonstrates the match between these techniques, and evidences that they provide distinct point of views.

The possibility to investigate many structural and dynamical aspects of chiral Yb-complexes is especially relevant for the study of asymmetric catalysis, where the structural changes, as well as the ligand dynamics, play a key role in the catalytic activity. Here we tested these techniques on enantioselective catalytically active complexes and on several potentially active catalysts, which offers a rich series of cases, making these results of general interest. In principle, any Yb-containing system can be studied, including biomolecules where Yb can be introduced through especially designed tags, or replacing the Ca(II) ion in calcium binding metalloproteins.

Related publications

DI BARI, L.; LELLI, M.; PINTACUDA, G.; SALVADORI, P., "Yb(fod)₃ in the Spectroscopic Determination of the Configuration of Chiral Diols: A Survey of the Lanthanide Diketonate Method", *Chirality* **14** (2002), 265–273.

DI BARI, L.; LELLI, M.; PINTACUDA, G.; PESCITELLI, R.; MAR-CHETTI, F.; SALVADORI, P., "Solution versus Solid-State Structure of Ytterbium Heterobimetallic Catalysts", *J. Am. Chem. Soc.* **125** (2003), 5549–5558.

DI BARI, L.; LELLI, M.; SALVADORI, P., "Ligand Lability and Chirality Inversion in Yb Heterobimetallic Catalysts", *Chem. Eur. J.* **10** (2004), 4594–4598.

LELLI, M.; PINTACUDA, G.; CUZZOLA, A.; DI BARI, L., "Monitoring Proton Exchange and Solution Conformation of Chiral Ytterbium Complexes trough Near-IR CD", *Chirality* **17** (2005), 201–211.

LELLI, M.; DI BARI, L.; SALVADORI, P., "Solution Study of Chiral 1,2-Diol Yb³⁺ Chelates through Near-IR Circular Dichroism", *Tetrahedron: Asymmetry* **18** (2007), 2876–2885.

LELLI, M.; DI BARI, L.; SALVADORI, P., "Dimerization and Structural Rearrangement of Yb Macrocyclic Complex Investigated in Solution through NMR and NIR CD", in preparation.

LELLI, M.; PASTORE, C.; DI BARI, L.; SALVADORI, P., "Solution Structure, Conformational Equilibria, and Axial-Coordination Dynamics of [Yb(THED)] Complex Investigated through Paramagnetic NMR", in preparation.

Appendices

Appendix A
Experimental session

A.1. General procedures, instruments and materials

NMR spectra were recorded on Varian VXR 300 and on VXR 600 spectrometers operating at 7 and 14 T, respectively, and equipped with a VT unit stable within 0.1 °C. Where not otherwise specified, the temperature was 298 K. Standard pulse sequences were used. All ^{1}H and ^{13}C NMR shifts are referred to TMS.

NIR absorption spectra were recorded at room temperature on a PERKIN-ELMER Lambda 19 UV/VIS/NIR spectrophotometer, under the following conditions: scan speed 7.5 nm/min, bandwidth 1.0 nm, NIR sensitivity 3, smooth bandwidth 2 nm. A square 1 cm quartz cell was used. NIR CD spectra were recorded on a JASCO 200 D spectropolarimeter, operating between 750 and 1350 nm, modified with a tandem Si/InGaAs detector with dual photomultiplier amplifier [1].

UV-Vis absorption spectra were recorded using UV-VIS Varian CARY 4E spectrophotometer at room temperature using a cylindrical quartz cell with 0.01 cm pathlength.

UV-Vis CD spectra were recorded on a JASCO J 600 spectropolarimeter, at room temperature under the following conditions: path length 0.1 cm, band-passing 1.0 nm, response 4 sec, speed 20 nm/min.

ESI MS spectra were acquired on a Perkin-Elmer Sciex API III plus triple quadrupole mass spectrometer equipped with an API ion source and an articulated ionspray interface. The ESI spectra were obtained under the following experimental conditions: ionspray voltage, 5.5 kV; orifice voltage, 35 V; scan range, m/z 200-1500; resolution above 1 Da; scan rate as appropriate. The nature of the observed species was assigned on the basis of the isotopic pattern and from the fragmentation profile in MS-MS experiments. ESI product ions were produced by collision-induced decomposition (CID) of selected precursor ions in the collision cell of the PE Sciex API III plus and mass-analyzed using the second analyser of the instrument. Other experimental conditions for the CID included:

collision energy 20 eV; collision gas thickness, 250 10^{13} molecules/cm^2; scan range was variable, depending on the m/z values of the precursor ion.

Dry THF was obtained by distilling the commercial product (BACKER) under N_2, on Na-K alloy. Commercial triethyl orthoformate was purified by distillation; dry acetonitrile, DMSO, EtOH, CH_2Cl_2, n-hexane, and n-pentane were purified from the commercial product (FLUKA) using the routine procedures. Commercial Yb(OTf)$_3$ was used (FLUKA). LuCl$_3$ was purchased from ALFA. (S)-1,1′-bis(2-naphthol) was resolved from the commercial FLUKA racemate following the literature procedure [2]. t-BuONa was prepared by refluxing t-BuOH on Na in inert atmosphere for 3 days; the t-BuOH excess was removed at low pressure. KHMDS (Potassium bis-(trimethylsilil)-azide) was used as the SIGMA-ALDRICH toluene solution 0.5 M. Cyclen was purchased from SIGMA-ALDRICH.

A.2. Experimental details of Chapter 4

A.2.1. Instrumental details

NMR: all the spectra were recorded dissolving the sample in commercial anhydrous d_8-THF. The NMR spectra investigating the homochiral exchange between K$_3$[Yb((R)-BINOL)$_3$] and (R)-BINOL-H$_2$ (EXSY, build-up and other 1D experiments) were measured on a sample of K$_3$[Yb((R)-BINOL)$_3$] 1.83 mM and with 1 equivalent of (R)-BINOL-H$_2$. The NMR spectra for studying the heterochiral exchange between K$_3$[Yb((R)-BINOL)$_3$] and (S)-BINOL-H$_2$ (EXSY, and other 1D experiments) were measured on a sample of K$_3$[Yb((R)-BINOL)$_3$] 11.0 mM and (S)-BINOL-H$_2$ 33.0 mM.

NIR absorption spectra were acquired in dry THF, at 298 K; Na$_3$[Yb((S)-BINOL)$_3$] 12.6 mM; K$_3$[Yb((S)-BINOL)$_3$] 19.4 mM.

NIR CD spectra were acquired in dry THF, at 298 K and 193 K; Na$_3$[Yb((S)-BINOL)$_3$] 12.6 mM; K$_3$[Yb((S)-BINOL)$_3$] 19.4 mM. The bandwidth was 2.4 nm and further narrowing of slit did not improve resolution. The spectra were recorded with 8 and 16 acquisitions at 50 nm/min with 0.5 s time-constant, and using 1 cm-quartz cell. The heterochiral exchange between K$_3$[Yb((R)-BINOL)$_3$] and (S)-BINOL-H$_2$ was measured on a sample of K$_3$[Yb((R)-BINOL)$_3$] 11.0 mM with (S)-BINOL-H$_2$ 33.0 mM.

A.2.2. Synthesis of Na$_3$[Yb((S)-BINOL)$_3$]

To a stirred solution of (S)-1,1′-bis(2-naphthol) (258 mg, 0.9 mmol) in 8.0 mL THF, were added in the order: a solution of Yb(OTf)$_3$(186 mg, 0.3 mmol) in 3.0 mL THF, t-BuONa (173 mg, 1.8 mmol) and a solution of 5.4 μL (0.3 mmol) of H_2O dissolved in 0.3 mL THF, at room temperature and under argon atmosphere. After 1h, the stirring was stopped and the clear yellow solution was separated by decantation. The volatile components were removed, the complex was dissolved in 1.9 mL of THF, and this solution was filtered over a 0.2 μm PTFE filter. n-Pentane (0.3 mL) was added dropwise to 0.6 mL of the solution, and the complex slowly crystallised over 1 day as pale yellow crystals.

^{1}H NMR (d_8-THF): δ 43.7 (br s, 6H, H-3), 14.5 (s, 6H, H-4), 7.4 (d, 6H, $J = 8$ Hz, H-5), 4.5 (m, 6H, H-6), 3.2 (m, 6H, H-7), −2.3 (d, 6H, $J = 8$ Hz, H-8). ^{13}C NMR (d_8-THF): δ 149.4, 132.7, 125.1, 119.0, 118.2, 117.2.

A.2.3. Synthesis of K₃[Yb((S)-BINOL)₃] and K₃[Yb((R)-BINOL)₃]

To a stirred solution of (S)-1,1′-bis(2-naphthol) (258 mg, 0.9 mmol); in 8.0 mL THF, were added in the order: a solution of Yb(OTf)₃ (186 mg, 0.3 mmol) in 3.0 mL THF, KHMDS (3.6 mL, 1.8 mmol, 0.5 M in toluene) and a solution of 5.4 μL (0.3 mmol) of H_2O dissolved in 0.3 mL THF, at room temperature and under argon atmosphere. After 1h, the stirring was stopped and the clear yellow solution was separated by decantation. The volatile components were removed, the complex was dissolved in 1.9 mL of THF, and this solution was filtered over a 0.2 μm PTFE filter. n-Pentane (0.3 mL) was added dropwise to 0.6 mL of the solution, and the complex slowly crystallised over 1 day as pale yellow crystals. To prepare K₃[Yb((R)-BINOL)₃] the same procedure was used but starting from (R)-1,1′-bis(2-naphthol) instead of (S)-1,1′-bis(2-naphthol).

^{1}H NMR (d_8-THF): δ 20.5 (br s, 6H, H-3), 10.1 (s, 6H, H-4), 7.5 (d, 6H, H-5), 5.7 (m, 6H, H-6), 5.4 (m, 6H, H-7), 3.0 (d, 6H, H-8). ^{13}C NMR (d_8-THF): δ 132.0, 129.0, 126.5, 122.1, 121.8, 119.6.

A.2.4. Synthesis of Na₃[Lu((S)-BINOL)₃]

To a stirred solution of (S)-1,1′-bis(2-naphthol) (143 mg, 0.5 mmol) in 1.8 mL THF, were added in the order: a solution of dry LuCl₃ (70.3 mg, 0.25 mmol) in 2.8 mL THF, t-BuONa (168 mg, 1.75 mmol) and a solution of 50 μL (2.8 mmol) of H_2O dissolved in 0.3 mL THF, at room temperature and under argon atmosphere. The mixture was stirred for 20 h at 50 °C. The clear solution was separated by decantation. The volatile components were removed under vacuum (20 mmHg), the complex was dissolved in 1.9 mL THF and the solution was filtered over a 0.2 μm PTFE filter. n-Pentane (0.3 mL) was added dropwise to 0.6 mL of the solution, and the complex slowly crystallised over 1 day as colorless crystals.

^{1}H NMR (d_8-THF): δ 6.7 (m, 6H, H-8), 6.8 (m, 12H, H-6 and H-7), 7.6 (d, 6H, $J = 9$ Hz, H-4), 7.5 (d, 6H, $J = 9$ Hz, H-5), 7.4 (d, 6H, $J = 9$ Hz, H-3). ^{13}C NMR (d_8-THF): δ 128.3, 128.3, 127.1, 125.9, 124.7, 120.2.

A.2.5. Crystal structure determination

A colorless hexagonal prism of $Na_3[Yb((S)\text{-BINOL})_3]\cdot6THF$, of 0.33 × 0.24 × 0.20 mm dimensions, was sealed in a glass capillary and mounted on a Bruker P4 diffractometer, equipped with a graphite-monochromated Mo-K_α radiation ($\lambda = 0.71073$ Å). The cell parameters, calculated from the setting angles of 40 reflections having $7.54° < \theta < 12.06°$, are listed in Table A.1, together with some other structural details. A total of 5672 intensities between $2.1° < \theta < 25.0°$ was collected with the $\omega/2\theta$ scan mode. Three standard reflections were measured every 97 measurements to check the sample decay and equipment stability. The intensities were corrected for Lorentz and polarization effects and for absorption by means of a Gaussian method based on the crystal shape [3]. The equivalent reflections were merged, leaving unmerged the Friedel pairs. 2459 $F_o's^2$, including the Friedel pairs, were then obtained (R_{int}, $\Sigma|F_o^2 - F_o^2{}_{(mean)}|/\Sigma[F_o^2] = 0.0502$).

The structure solution was obtained in the space group P 6_3 by the standard Patterson and Fourier methods. The refinements, based on full-matrix least-squares on F^2, were done by means of SHELX97 program [4]. The H atoms were in part located in the difference Fourier map and in part placed in calculated positions. They were refined with the riding constraints. The absolute configuration was assumed to be correct on the basis of the Flack index −0.02(3) [5]. The final refinement cycles gave the reliability factors listed in Table A.1. The final difference Fourier map showed three peaks among 0.38 and 0.58 $e\cdot$Å^{-3} at distances between 1.01 and 1.16 Å from the Yb atom. They are probably due to a defective description of its thermal motion and of absorption effects. The other residual electron-density peaks are lower than 0.3 $e\cdot$Å^{-3} and none of them can be attributed to a water oxygen connected to ytterbium.

A.3. Experimental details of Chapter 5

A.3.1. Instrumental details

NMR: the spectra were recorded dissolving the sample in commercial D_2O, CD_3OD, CD_3CN, d_6-DMSO, and H_2O. Where not otherwise specified, the NMR spectra of [Yb((S)-THP)] in D_2O and H_2O were acquired on a sample 71 mM concentrated. The concentrations in the other solvents ranged from 20 to 80 mM. The NMR spectra of the dimeric

Table A.1. Crystallographic Data for $Na_3[Yb((S)\text{-}BINOL)_3]\cdot 6THF$.

Empirical formula	$C_{84}H_{84}Na_3O_{12}Yb$
Formula weight	1527.52
Temperature, K	298(2)
Crystal system	Hexagonal
Space group	$P6_3$ (No. 173)
a, Å	15.286(1)
b, Å	15.286(1)
c, Å	18.709(2)
α, deg	90
β, deg	90
γ, deg	120
Volume, Å^3	3785.9(5)
Z	2
ρ_{calc}, Mg/m^3	1.340
μ, mm^{-1}	1.313
Observed reflection $[I > 2\sigma(I)]$	1604
Data / restraints / parameters	2459 / 1 / 276
$R(F_o)[I > 2\sigma(I)]$	0.0395
$Rw(F_o^2)$	0.0801

$R(F_o) = \Sigma||F_o| - |F_c||/\Sigma|F_o|$; $Rw(F_o^2) = [\Sigma[w(F_o^2 - F_c^2)^2]/\Sigma[w(F_o^2)^2]]^{1/2}$; $w = 1/[\sigma^2(F_o^2) + (0.0427Q)^2]$ where $Q = [MAX(F_o^2, 0) + 2F_c^2]/3$.

$[Yb((S)\text{-}THP)]_2$ species (40 mM) were acquired in CD_3CN, d_6-DMSO. The NMR spectra of [Yb(THED)] were acquired in CD_3OD, CD_3CN, and d_6-DMSO, with concentration ranging from 50 to 80 mM.

NIR absorption spectra were acquired at 298 K, on $[Yb((S)\text{-}THP)]$ (27 mM in H_2O, and 80 mM in CH_3CN), and on $[Yb((S)\text{-}THP)]_2$ 40 mM in CH_3CN.

NIR CD spectra were acquired at 298 K, on $[Yb((S)\text{-}THP)]$ (27 mM in H_2O, and 80 mM in CH_3CN), and on $[Yb((S)\text{-}THP)]_2$ 40 mM in CH_3CN. The bandwidth was 3.4 nm and further narrowing of slit did not improve resolution. The spectra were recorded with 4 acquisitions at 5-50 nm/min with 8-0.5 s time-constant, respectively. A 1 cm-quartz cell was used.

A.3.2. Synthesis of ligand 1,4,7,10-tetrakis((S)-2-hydroxypropyl)-1,4,7,10-tetraazacyclododecane ((S)-THP)

Homochiral (S)-THP was synthesized by addition of enantiopure (S)-propylen oxide (FLUKA) to the tetraazamacrocyclic cyclen ring (1,4,7,10-tetraazacyclododecane), as reported in the literature [6]. The reaction was followed through ESI MS spectroscopy until the complete disappearance of the mass peaks of the cyclen and of all the partially substituted species. The ligand was purified through crystallization from dry acetonitrile by addition of n-hexane.

^{1}H NMR (CDCl$_3$, 298 K, 7 T): δ 1.1 (12H, d, $J = 6$ Hz, $-$CH$_3$), 1.7 (4H, s, $-$OH), 2.0-2.3 (16H, m, NCH$_2$CH$_2$N and $-$CH$_2$CH(OH)CH$_3$), 2.9 (8H, d, $J = 9$ Hz, NCH$_2$CH$_2$N), 3.9 (4H, m, CH(OH)CH$_3$).

^{13}C NMR (CDCl$_3$, 298 K, 7 T): δ 20.0, 51.3, 62.8, 63.7.

A.3.3. Synthesis of the complex [Yb((S)-THP)](TfO)$_3$

The ytterbium complex was prepared mixing the ligand (S)-THP with anhydrous Yb(OTf)$_3$ in dry acetonitrile and in presence of triethyl orthoformate as drying agent, following the literature procedure reported for the complex of La, Eu, and Lu [6]. The complex [Yb((S)-THP)](TfO)$_3$ was purified by crystallization from acetonitrile by addition of CH$_2$Cl$_2$ obtaining a white power.

^{1}H NMR (H$_2$O, 298 K, pH $= 0.79$, 14 T): δ 52.1 (4H, s, axial NCH$_2$CH$_2$N), 12.5 (4H, s, equatorial NCH$_2$CH$_2$N), 8.8 (4H, s, equatorial NCH$_2$CH$_2$N), -4.0 (12H, s, CH$_3$), -6.6 (4H, s, CH$_2$CH(OH)CH$_3$), -16.6 (4H, s, axial NCH$_2$CH$_2$N), -17.1 (4H, s, CH(OH)CH$_3$), -28.6 (4H, s, CH$_2$CH(OH)CH$_3$). ^{13}C NMR (H$_2$O, 298 K, pH $= 0.79$, 14 T): δ 69.9, 45.9, 24.6, 11.9, 9.0.

A.3.4. Synthesis of the complex [Yb(THED)](TfO)$_3$ (THED is 1,4,7,10-tetrakis-(2-hydroxyethyl)-1,4,7,10-tetraazacyclododecane)

The THED ligand was synthesized from cyclen with exhaustive addition of ethylene oxide, following the procedure reported in the literature [7]. The reaction was followed with ESI MS to completeness. The ligand was purified through crystallization from dry acetonitrile by addition of n-hexane. The ytterbium complex was prepared mixing the ligand THED with anhydrous Yb(OTf)$_3$ in dry acetonitrile and in presence of triethyl orthoformate as drying agent, following the literature procedure reported for analogous lanthanide complexes [7]. The complex was purified through crystallization from acetonitrile by addition of CH$_2$Cl$_2$.

^{1}H NMR (THED, CDCl$_3$, 298 K, 7 T): δ 2.6 (16H, s, NCH$_2$CH$_2$N), 2.5 (8H, s, CH$_2$CH$_2$OH), 3.6 (8H, s, CH$_2$CH$_2$OH).

^{1}H NMR ([Yb(THED)], CD$_3$CN, 298 K, 14 T): δ 111.7 (4H, s, axial NCH$_2$CH$_2$N), 27.7 (4H, s, equatorial NCH$_2$CH$_2$N), 23.5 (4H, s, equatorial NCH$_2$CH$_2$N), −20.4 (8H, s, CH$_2$CH$_2$OH, CH$_2$CH$_2$OH), −23.9 (4H, s, axial NCH$_2$CH$_2$N), −33.1 (4H, s, CH$_2$CH$_2$OH), −60.5 (4H, s, CH$_2$CH$_2$OH).

A.4. Experimental details of Chapter 6

A.4.1. Instrumental details

NMR: the spectra of the diol titration of [Yb(fod)$_3$] were acquired in CDCl$_3$ using a sample about 0.1 M concentrated. The spectra of the Yb(OTf)$_3$/diol mixtures were acquired in CD$_3$CN with 20 mM of Yb(OTf)$_3$ concentration.

UV-Vis CD spectra were acquired at 298 K using [Yb(fod)$_3$] 0.35 mM and 0.18 mM in CCl$_4$, path length 0.1 cm, band-passing 1 nm, time-constant 4 s, scan speed 20 nm/min.

NIR CD spectra of the [Yb(fod)$_3$]/diol mixtures were acquired at 298 K, using [Yb(fod)$_3$] 5 mM in CCl$_4$. The path length was 10 cm, for each spectrum 4 acquisitions at the rate of 50 nm/min, 0.5 s time-constant, and 3.2 nm band-passing were used. NIR CD spectra of the Yb(OTf)$_3$/diol mixtures were acquired at concentrations ranging from 5.25 to 60 mM in CH$_3$CN and CH$_3$NO$_2$. Spectra were recorded with a number of scans varied from 4 to 32 depending on the signal to noise (typically 8 scansions per spectrum, were not otherwise specified), 3.2 nm bandwidth, time-constant 0.5 s (with scan speed 50 nm/min), 1.0 s and 2.0 s (with scan speed 20 nm/min); 1.0 cm quartz cell was used for samples in CH$_3$CN and 10.0 cm quartz cell for samples in CH$_3$NO$_2$.

A.4.2. Synthesis of the used chiral diol

Enantiopure (R,R) and (S,S)-2,3-buthane diol (**9**) (e.e. 99%), and the diols **11** (e.e. 99%), and **13** (e.e. 99%) are commercially available and purchased from SIGMA-ALDRICH and FLUKA. The diols **10** (e.e. 96%), **12** (e.e. 99%), **14** (e.e. 70%), **16** (e.e. 99%), **17** (e.e. 99%), **18** (e.e. 30%), **19** (e.e. 56%), **20** (e.e. 89%) (see Figure 6.6) were synthesized from the corresponding commercial available alkenes (SIGMA-ALDRICH and FLUKA) following the Sharpless procedure [8], as reported in litera-ture [9]. The diol **15** was synthesized according the literature procedure [10], with a global yield of 53% starting from 3,5-dimetoxy benzoic acid. The product was isolated with an e.e. 68% determined by HPLC (peaks at 63.7 and 75.3 min, $\lambda = 220$ nm, flow 0.5 ml/min), using a CHIRAL-PAK AD column in hexane/iPrOH 9/1. Mp 95-105 °C. $[\alpha]_D^{25} = -0.274$ (c 0.98; EtOH); $[\alpha]_{577}^{25} = -0.282$ (c 0.98; EtOH). Spectroscopic data (^{1}H NMR, ^{13}C NMR, IR, MS) are in agreement with the literature [10].

A.4.3. Analysis of the curves of the NIR CD titration of Yb(OTf)₃ with the diols 9 and 13

The mathematical expressions used to fit the experimental data in titration of Yb(OTf)₃ with the diols **9** and **13** (Figure 6.16 and 6.18) were derived considering the simultaneous action of all the chemical equilibria described in the equations (6.6-8). The three concentration constants K_1, K_2, K_3 were considered as (A.1). By assuming unitary activity coefficients and taking as a reference the solutions with 1 M concentration, K_1, K_2, K_3 are equal to the corresponding thermodynamic stability equilibrium constants

$$\frac{[Yb(d)]}{[Yb][d]} = K_1, \quad \frac{[Yb(d)_2]}{[Yb(d)][d]} = K_2, \quad \frac{[Yb(d)_3]}{[Yb(d)_2][d]} = K_3 \quad (A.1)$$

where $[Yb(d)]$, $[Yb(d)_2]$, $[Yb(d)_3]$, $[d]$, $[Yb]$ are the molar concentrations of the mono-, di-, tris-diol coordinated species, the free diol and the free metal, respectively. From the combination of all these equilibria it results

$$[d]^4 + (K_3^{-1} + C(3-\alpha))[d]^3 + K_3^{-1}(K_2^{-1} + C(2-\alpha))[d]^2$$
$$+ K_3^{-1}K_2^{-1}(K_1^{-1} + C(1-\alpha))[d] + K_3^{-1}K_2^{-1}K_1^{-1}C\alpha = 0 \quad (A.2)$$

and

$$[Yb(d)_3] = [d]^3C/([d]^3 + K_3^{-1}[d]^2 + K_3^{-1}K_2^{-1}[d] + K_3^{-1}K_2^{-1}K_1^{-1}) \quad (A.3)$$

$$[Yb(d)_2] = K_3^{-1}[d]^2C/([d]^3 + K_3^{-1}[d]^2 + K_3^{-1}K_2^{-1}[d] + K_3^{-1}K_2^{-1}K_1^{-1}) \quad (A.4)$$

$$[Yb(d)] = K_3^{-1}K_2^{-1}[d]C/([d]^3 + K_3^{-1}[d]^2 + K_3^{-1}K_2^{-1}[d] + K_3^{-1}K_2^{-1}K_1^{-1}) \quad (A.5)$$

where C is the total metal concentration and α are the equivalents of added diol. The numerical solution of (A.2) allows one to calculate the concentration of the single species through (A.3-5). These are in turn related to the observed CD intensity (ΔA) through

$$\Delta A = \Delta\varepsilon_1[Yb(d)] + \Delta\varepsilon_2[Yb(d)_2] + \Delta\varepsilon_3[Yb(d)_3] \quad (A.6)$$

where $\Delta\varepsilon_1$, $\Delta\varepsilon_2$, $\Delta\varepsilon_3$ are the molar coefficients of the single chiral species. All the equations (A.2-6) were included into a computer program (written in *Mathematica* language) [11], which optimizes the three binding constants K_1, K_2, K_3 and of the three coefficients $\Delta\varepsilon_1$, $\Delta\varepsilon_2$, $\Delta\varepsilon_3$ through a least-square fitting of the experimental CD intensities (ΔA).

References

[1] CASTIGLIONI, E., *Book f Abstract, 6th international Conference on CD*; Pisa, September **1997**.

[2] CAI, D. D. L.; HUGHES, T. R.; VERHOEVEN, P. J. R., *Tetrahedron lett.* **1995**, *36*, 7991.

[3] XSCANS, *X-ray Single Crystal Analysis System*, rel. 2.1 Siemens Analytical X-ray Instruments Inc., Madison, WI, USA, **1996**.

[4] G. M. SHELDRICK, SHELXTL-Plus, Rel. 5.1, Bruker Analytical X-Ray Instruments Inc., Madison, WI, USA, **1997**.

[5] FLACK, H. D., *Acta Cryst.* **1983**, *A39*, 876–881.

[6] HANCOCK, R. D.; SHAIKJEE, M. S.; DOBSON, S. M.; BOEYENS, J. C. A., *Inorg. Chim. Acta* **1988**, *154*, 229–238.

[7] MORROW, J. R.; CHIN, K. A. O., *Inorg. Chem.* **1993**, *32*, 3357–3361.

[8] (a) KOLB, H. C.; VAN NIEUWENHZE, M. S.; SHARPLESS, K. B., *Chem. Rev.* **1994**, *94*, 2483; (b) SALVADORI, P.; PINI D.; PETRI A., *Synlett.* **1999**, 1181; (c) ROSINI, C.; SUPERCHI, S.; DONNOLI, M. I., *Enantiomer*, **1999**, *4*, 3–23.

[9] DI BARI, L.; PESCITELLI, G.; PRATELLI, C.; PINI, D.; SALVADORI, P., *J. Org. Chem.* **2001**, *66*, 4819–4825.

[10] SALVADORI, P.; SUPERCHI, S.; MINUTOLO, F., *J. Org. Chem.* **1996**, *61*, 4190–4191, and references therein.

[11] *Mathematica* 5.0, Wolfram Research, Inc.

Appendix B
List of Abbreviations

(...)	Numbered equations
[...]	Numbered chemical equilibria
acac	acethylacetonate
BINOL	1,1'-bis(2-naphtholate)
BINOL-H$_2$	2-binaphthol
CD	Circular Dichroism
CN	Coordination Number
Cyclen	1,4,7,10-tetraazacyclododecane
DCTA	1,2-diaminocyclohexane-N,N,N',N'-tetraacetic acid
dmp	dipivalomethanate
DMSO	dimethylsulfoxide
DMTA	dimolibdenum tetraacetate [Mo$_2$(AcO)$_4$]
DN	Donor Number
DOTA	1,4,7,10-tetraazacyclododecane-1,4,7,10-tetraacetic acid
DOTAM	1,4,7,10-tetrakis(2-carbamoilethyl)-1,4,7,10-tetraazacyclododecane
DOTAMNp	1,4,7,10-tetrakis[(S)-1-(1-naphthyl)ethylcarbamoilmethyl]-1,4,7,10-tetraazacyclododecane
DOTAMPh	1,4,7,10-tetrakis[(R)-1-(phenyl)ethylcarbamoilmethyl]-1,4,7,10-tetraazacyclododecane
DOTMA	(1R,4R,7R,10R)-$\alpha,\alpha',\alpha'',\alpha'''$-tetramethyl-1,4,7,10-tetraazacyclododecane-1,4,7,10-tetraacetic acid
DOTPBz$_4$	1,4,7,10-tetraazacyclododecane-1,4,7,10-tetrakis(methylenebenzylphosphinic acid)
DTPA	diethylenetriamine-N,N,N',N',N''-pentaacetic acid
EDTA	ethylen diamine tetraacetate
ESI MS	ElectroSpray Ionization Mass Spectrometry
EXSY	EXchange SpectroscopY

FAB MS	Fast Atom Bombardment Mass Spectrometry
fod	6,6,7,7,8,8,8-heptafluoro-2,2-dimethyl-3,5-octanedioate
GF	Geometrical Factor
HEDTA	N'-hydroxyethylendiamine-N,N,N'-triacetic acid
HMQC	Heteronuclear Multiple-Quantum Coherence
HSAB	Hard-Soft Acids and Bases (Pearson's terminology)
ICD	Induced Circolar Dichroism
***i*PrO**	*iso*-propoxide
LDI-TOF MS	Laser Desorption Ionization Time-Of-Flight Mass Spectrometry
Ln	lanthanide
M₃[Ln(BINOL)₃]	Heterobimetllic complexes (Ln = lanthanide ion, M = alkaline cation, BINOL as a ligand)
MM2	N. L. Allinger's MM2 force field
MO	Molecular Orbital
MRI	Magnetic Resonance Imaging
NIR CD	Near-InfraRed Circular Dichroism
NO₂BnDOTMA	$(1R,4R,7R,10R)$-$\alpha,\alpha',\alpha'',\alpha'''$-tetramethyl-[(S)-2-(nitrobenzyl)]-1,4,7,10-tetraazacyclododecane-1,4,7,10-tetra-acetic acid
PCS	PseudoContact Shift
PERSEUS	Paramagnetic Enhanced Relaxation and Shift for Eliciting the Ultimate Structure
SA	Square Antiprism geometry
***t*BuO**	*tert*-butoxide
TfO⁻	trifluoromethanesulfonate anion (triflate anion)
THED	1,4,7,10-tetrakis-(2-hydroxyethyl)-1,4,7,10-tetraazacyclododecane
THF	tetrahydrofuran
THP	1,4,7,10-tetrakis-((S)-2-hydroxypropyl)-1,4,7,10-tetraazacyclododecane
TSA	Twisted-Square Antiprism geometry
XRD	X-Ray Diffraction

Appendix C
List of Numbered Molecules

1: Ln = Yb, M = Na
2: Ln = Yb, M = K
3: Ln = Yb, M = Li
4: Ln = Lu, M = Na

Heterobimetallic M₃[Ln(BINOL)₃] complexes.

(S)-THP

5

THED

6

[Yb(dpm)₃]

7

[Yb(fod)₃]

8

9

10

11

12

13

14

15

16

17

18

19

20

THESES

This series gathers a selection of outstanding Ph.D. theses defended at the Scuola Normale Superiore since 1992.

Published volumes

1. F. COSTANTINO, *Shadows and Branched Shadows of 3 and 4-Manifolds*, 2005. ISBN 88-7642-154-8

2. S. FRANCAVIGLIA, *Hyperbolicity Equations for Cusped 3-Manifolds and Volume-Rigidity of Representations*, 2005. ISBN 88-7642-167-x

3. E. SINIBALDI, *Implicit Preconditioned Numerical Schemes for the Simulation of Three-Dimensional Barotropic Flows*, 2007. ISBN 978-88-7642-310-9

4. F. SANTAMBROGIO, *Variational Problems in Transport Theory with Mass Concentration*, 2007. ISBN 978-88-7642-312-3

5. M. R. BAKHTIARI, *Quantum Gases in Quasi-One-Dimensional Arrays*, 2007. ISBN 978-88-7642-319-2

6. T. SERVI, *On the First-Order Theory of Real Exponentiation*, 2008. ISBN 978-88-7642-325-3

7. D. VITTONE, *Submanifolds in Carnot Groups*, 2008. ISBN 978-88-7642-327-7

8. A. FIGALLI, *Optimal Transportation and Action-Minimizing Measures*, 2008. ISBN 978-88-7642-330-7

9. A. SARACCO, *Extension Problems in Complex and CR-Geometry*, 2008. ISBN 978-88-7642-338-3

10. L. MANCA, *Kolmogorov Operators in Spaces of Continuous Functions and Equations for Measures*, 2009. ISBN 978-88-7642-336-9

11. M. LELLI, *Solution Structure and Solution Dynamics in Chiral Ytterbium(III) Complexes*, 2009. ISBN 978-88-7642-349-9

Volumes published earlier

H.Y. FUJITA, *Equations de Navier-Stokes stochastiques non homogènes et applications*, 1992.

G. GAMBERINI, *The minimal supersymmetric standard model and its phenomenological implications*, 1993. ISBN 978-88-7642-274-4

C. DE FABRITIIS, *Actions of Holomorphic Maps on Spaces of Holomorphic Functions*, 1994. ISBN 978-88-7642-275-1

C. PETRONIO, *Standard Spines and 3-Manifolds*, 1995.
ISBN 978-88-7642-256-0

I. DAMIANI, *Untwisted Affine Quantum Algebras: the Highest Coefficient of* det H_η *and the Center at Odd Roots of 1*, 1996.
ISBN 978-88-7642-285-0

M. MANETTI, *Degenerations of Algebraic Surfaces and Applications to Moduli Problems*, 1996. ISBN 978-88-7642-277-5

F. CEI, *Search for Neutrinos from Stellar Gravitational Collapse with the MACRO Experiment at Gran Sasso*, 1996. ISBN 978-88-7642-284-3

A. SHLAPUNOV, *Green's Integrals and Their Applications to Elliptic Systems*, 1996. ISBN 978-88-7642-270-6

R. TAURASO, *Periodic Points for Expanding Maps and for Their Extensions*, 1996. ISBN 978-88-7642-271-3

Y. BOZZI, *A study on the activity-dependent expression of neurotrophic factors in the rat visual system*, 1997. ISBN 978-88-7642-272-0

M.L. CHIOFALO, *Screening effects in bipolaron theory and high-temperature superconductivity*, 1997. ISBN 978-88-7642-279-9

D.M. CARLUCCI, *On Spin Glass Theory Beyond Mean Field*, 1998.
ISBN 978-88-7642-276-8

G. LENZI, *The MU-calculus and the Hierarchy Problem*, 1998.
ISBN 978-88-7642-283-6

R. SCOGNAMILLO, *Principal G-bundles and abelian varieties: the Hitchin system*, 1998. ISBN 978-88-7642-281-2

G. ASCOLI, *Biochemical and spectroscopic characterization of CP20, a protein involved in synaptic plasticity mechanism*, 1998.
ISBN 978-88-7642-273-7

F. PISTOLESI, *Evolution from BCS Superconductivity to Bose-Einstein Condensation and Infrared Behavior of the Bosonic Limit*, 1998.
ISBN 978-88-7642-282-9

L. PILO, *Chern-Simons Field Theory and Invariants of 3-Manifolds*,1999.
ISBN 978-88-7642-278-2

P. ASCHIERI, *On the Geometry of Inhomogeneous Quantum Groups*, 1999. ISBN 978-88-7642-261-4

S. CONTI, *Ground state properties and excitation spectrum of correlated electron systems*, 1999. ISBN 978-88-7642-269-0

G. GAIFFI, *De Concini-Procesi models of arrangements and symmetric group actions*, 1999. ISBN 978-88-7642-289-8

N. DONATO, *Search for neutrino oscillations in a long baseline experiment at the Chooz nuclear reactors*, 1999. ISBN 978-88-7642-288-1

R. CHIRIVÌ, *LS algebras and Schubert varieties*, 2003.
ISBN 978-88-7642-287-4

V. MAGNANI, *Elements of Geometric Measure Theory on Sub-Riemannian Groups*, 2003. ISBN 88-7642-152-1

F.M. ROSSI, *A Study on Nerve Growth Factor (NGF) Receptor Expression in the Rat Visual Cortex: Possible Sites and Mechanisms of NGF Action in Cortical Plasticity*, 2004. ISBN 978-88-7642-280-5

G. PINTACUDA, *NMR and NIR-CD of Lanthanide Complexes*, 2004.
ISBN 88-7642-143-2

Fotocomposizione "CompoMat" Loc. Braccone, 02040 Configni (RI) Italy
Finito di stampare nel mese di febbraio 2009
BRAILLE-GAMMA s.r.l. Cittaducale (RI)